P9-CKD-913

A *Fly* FOR THE PROSECUTION

A *Fly* for the PROSECUTION

How Insect Evidence Helps Solve Crimes

M. Lee Goff

Harvard University Press
Cambridge, Massachusetts
London, England
2000

Printed in the United States of America

Drawings by Amy Bartlett Wright

Library of Congress Cataloging-in-Publication Data

Goff, M. Lee (Madison Lee)
A fly for the prosecution : how insect evidence helps solve crimes / M. Lee Goff.
p. cm.
Includes bibliographical references (p.) and index.
ISBN 0-674-00220-2 (alk. paper)
1. Forensic entomology. I. Title.

RA1063.45 .G64 2000
614'.1-dc21 99-058194

CONTENTS

A *Fly* FOR THE PROSECUTION

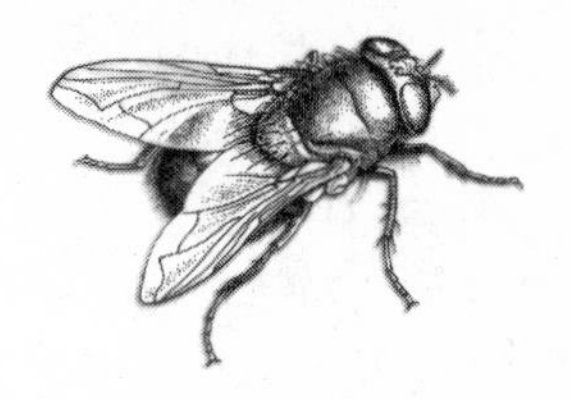

Prologue: Honolulu, 1984

It was a perfect morning for shoreline fishing and throwing nets for crabs. The sun was shining brightly and the air was perfumed with the scent of plumeria when the three fishermen set off for Pearl Harbor, only a few miles from home. At the abandoned Primo Brewery grounds, they parked and started on the short walk to the beach. As they went along the path, they noticed an unpleasant odor stronger than the smell of their bucket of bait. Peering over the fence in the direction of the stench, one spotted a dead body lying on its back.

When the homicide investigators arrived at the scene, they could see that the body was stretched across a shallow, brush-filled drainage ditch, with the head facing the ocean and the legs pointed inland toward Honolulu. The fingernails and toenails were painted bright red. The left arm was raised over the skull with a slight bend at the elbow, as if trying to defend against a blow. The left hand was missing, but the right hand was intact, although desiccated. The lower jaw had been separated from the skull and lay in the mud about 16 inches away. The left leg was crossed over the right. Three toes were missing from the left foot, but the legs were otherwise undisturbed. Many beetles and other insects were crawling both on and inside the body.

The body appeared to match the description of a woman who had been reported missing on September 9, 1984, some 19 days before the discovery of the corpse. When last seen alive, the woman, accompanied by a tall white male, was leaving a restaurant in Pearl City of which she was part owner. Her car was later found over 30 miles away, in the Waianae area. There was blood inside it.

The identity of the woman was established beyond reasonable doubt by dental x-rays. When reported missing, she had been wearing a black leotard with a white stripe along the side and a floral print skirt. By the time the body got to the morgue, all of the clothing had turned dark brown or black. Her head was almost completely stripped of flesh, and the exposed skull had been polished by the scraping mandibles of beetle larvae as they fed on the dried tissues. The rib cage was exposed, with some shreds of dried skin still clinging to it, and patches of parchment-like skin adhered to the neck and legs. The internal organs were missing. The only evidence of trauma the medical examiner could find was a fractured hyoid bone in the neck, consistent with manual strangulation. Now the police had an identification and a cause of death—homicide. But when did the victim die? Fortunately there were witnesses: the insects that were infesting the body. The only problem was how to get them to reveal their evidence to the investigators.

Having been called by the medical examiner, I arrived at the Honolulu morgue as the autopsy was being completed. Given the condition of the body, the procedure had not taken long. At the time, I had been actively involved in forensic entomology for only a little over a year, and the Honolulu Police Department and medical examiners were still getting used to the idea of an entomologist showing up at the morgue on a motorcycle with an insect net and a bag of vials. But my estimates of time of death had been helpful in resolving a couple of earlier cases, and on this occasion, I had been told I could bring along a graduate student, Marianne Early. She was in the final stages of her master's degree program in entomology and had been conducting decomposition studies on pig and cat carcasses on various parts of the island of Oahu. Up to this point, the medical examiner had regarded me as an isolated anomaly; now there were two of us.

What the body lacked in tissues, it made up for in insects. Marianne and I collected specimens of all the species of insects and of each stage of development of every species we could find and took them back to the laboratory at the University of Hawaii at Manoa in Honolulu for identification and analysis. The most obvious and numerous were the hide beetles and the maggots, the larvae of flies. There were three species of maggots on the body, in different locations and in different stages of development. I sorted each type into two sublots. I measured the length of each of the maggots in one of the lots, and used the average of these lengths to give me some idea of their stage of development. Then I preserved them in ethyl alcohol. I put the other sublot of maggots into a rearing chamber to complete their development to the adult stage.

Since most maggots look a lot alike, it is often difficult to identify them to the species level until they have metamorphosed into adults, which do look quite different from one another. Marianne and I had collected some relatively large maggots from the flesh remaining on the back of the body. From the shape of their mouthparts and the breathing openings, or spiracles, at the end of each maggot's body, I was able to tell that these were flesh flies, in the family Sarcophagidae, but could not identify the species until the maggots had completed their development into adult flies. There was also another type of maggot, somewhat smaller, on the back of the body. Over the next 2 weeks, we reared these maggots to adulthood, at which point we could tell they were a species of blow fly, *Phaenicia cuprina,* in the family Calliphoridae. The third type of maggot was a smaller fly in the family Piophilidae. These flies are commonly known as cheese skippers because they prefer to eat stored foods, especially cheese. The maggots of cheese skippers have a unique way of moving away from their food source—usually a corpse—before entering the pupal stage, where they will be transformed into adults. The maggot arches backward and grasps its anal papillae, the fleshy lobes protruding from the body near the anus, with its mouth hooks. Then the maggot flexes its muscles and releases its grasp, flinging itself into the air, a process called popping. Once safely away from the corpse or other food source, the maggot enters the pupal stage.

In addition to the maggots, we collected evidence of yet another species of fly that had fed on the body: empty pupal cases of another blow fly, *Chrysomya rufifacies,* were attached to the exposed ribs and caught in the folds of the skirt. We also found two types of beetles on the body. Hide beetles, in the family Dermestidae, were present both as adults and as larvae. These beetles normally feed on the dried skin of dead animals but may also feed on other dried, stored products that have a high protein content. The second kind of beetle was a species of checkered beetle, *Necrobia rufipes,* in the family Cleridae. Only a few adults of this species were present on the body.

By the time of this investigation, 1984, I had begun experimenting with computers to estimate the postmortem interval—the time elapsed between the death of a person and the discovery of the corpse—and had developed a computer program using data from the decomposition studies one of my graduate students and I had conducted. This was the first time I used the program in an actual criminal case. After entering all

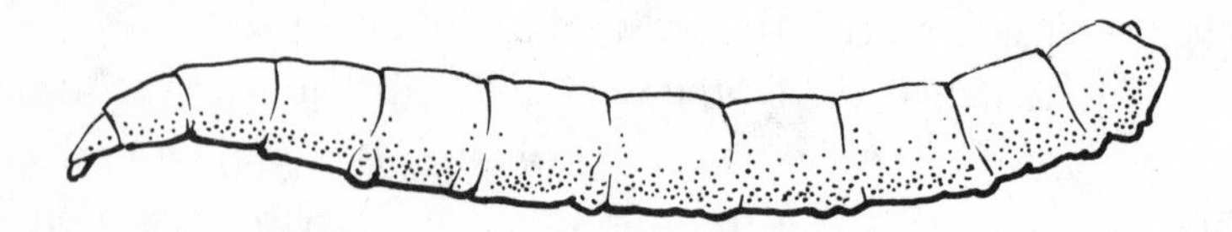

Larva of the cheese skipper *Piophila casei*

the data, I watched with distinct displeasure as my computer produced a completely illogical analysis. It seemed either that no such body existed, or that I had entered the data for two different bodies. Although this result was disconcerting, the test was a success of sorts. The program did detect a problem with the data; it just lacked the ability to solve it. I had simply made a few modifications to some off-the-shelf business software, and had allowed for only an either/or option. The resulting program was not capable of resolving the problem with the insects found on this particular body: sarcophagid (flesh fly) maggots should not have been on the body at the same time as empty pupal cases of the blow fly *Chrysomya rufifacies.* Normally both insects are present as larvae early in the decomposition process or both as pupae later in the process. The combination of sarcophagid larvae and empty puparial cases of *Chrysomya rufifacies* was not in any database available to the software program I was using. Another visit to the scene of the crime was in order. Late afternoon found me in a squad car with a detective and the medical examiner on our way to the drainage ditch in the Primo Brewery grounds.

At the site, we found that friends of the victim had already erected a wooden cross to commemorate her. By looking at the photographs of the scene, I pinpointed exactly where the body had lain across the shallow drainage ditch. Removing the brush from the surface of the ditch, I found water approximately 5 inches deep with a number of sarcophagid maggots moving across the surface. Here was the answer that had eluded the computer. Maggots can feed only on soft, moist flesh. As tissues lose moisture, they become more and more difficult for maggots to eat, until finally the maggots can no longer use the body as a food source. Since the victim's back had been partially submerged, these sarcophagid maggots had been able to continue feeding on the body far longer than they could have under dry conditions. Carefully examining the soil around the spot where the body had lain, I discovered some pupal cases of blow flies—the same kind of cases that would soon be formed by the blow fly maggots collected during the autopsy. I also collected some ants and a number of predatory beetles in the families Staphylinidae and Histeridae.

Blow fly adults of the species *Chrysomya rufifacies* can locate exposed human remains in a remarkably short period of time. In Hawaii, I have found them at test carrion less than 10 minutes after exposure of the carcass. Typically, the adults of this species of blow fly arrive at the body and feed briefly on blood and other secretions from the natural body openings or from wounds. The females then lay their eggs in dark areas either in body openings or underneath the body. Egg laying starts the biological clock that forensic entomologists use to estimate the postmortem interval. For *Chrysomya rufifacies,* egg laying can begin quite soon after the adult females reach the body and will continue, under Hawaiian conditions, for approximately the first 6 days following death. In the late summer to early fall in lowland areas on the island of Oahu, completion of development from egg to maggot to pupa and finally to adult usually requires 11 days. Since the only evidence of this species on the body was the empty pupal cases, discarded when the flies reach adulthood, I was confident that all *Chrysomya* maggots maturing on the body had completed development before it was discovered. Therefore, the minimum time since death was 17 days: 6 days of egg laying, followed by 11 days of development.

The cheese skipper maggots were still in the early stages of development, but in Hawaii, I have found, this fly typically does not invade remains until several days after death. The specimens I collected from the body were at the same stage of development as those I had collected from test animal carcasses set out for study after 19 days of decomposition.

The hide beetles also provided valuable clues for estimating the time of death. These beetles, which I identified as *Dermestes maculatus,* do not feed on moist tissues and arrive only when the remains have begun to dry. In lowland habitats on Oahu, they begin to arrive between 8 and 11 days after the onset of decomposition, and during decomposition studies I have gathered larvae comparable in size to those collected from this case beginning on day 19. The remaining species I collected, including the Histeridae and Staphylinidae found in the soil, were consistent with a postmortem interval of 19 to 20 days but did not yield more precise information.

Considering all the data, and having satisfied myself of the reason for the presence of the sarcophagid maggots, I determined that the most likely minimum postmortem interval was 19 days. This was the official estimate I gave to the medical examiner, Charles Odom.

In time a suspect was identified: the man in whose company the victim was last seen alive. I testified about the insect evidence first during a grand jury proceeding in April 1985. In late September 1985, I testified as to the probable time of death during a murder trial in the First Circuit Court in Honolulu.

The suspect was convicted of second degree murder and the major witnesses were flies. From that time on, I became a routine participant in investigations of decomposed human remains in Honolulu.

1

Beginnings

Anyone involved in death investigations quickly becomes aware of the connection between dead bodies and maggots. Insects are major players in nature's recycling effort, and in nature a corpse is simply organic matter to be recycled. Left to its own devices, nature quickly populates a corpse with a diverse community of organisms, all dedicated to reducing the body to its basic components. Very quickly, "worms" appear, crawling out of the various orifices. Until quite recently, most death investigators regarded these insects as merely a sign of decay, to be washed away or otherwise disposed of as quickly as possible, rather than potentially significant evidence. Thus while other forensic sciences, such as toxicology, forensic pathology, blood-spatter analysis, and ballistics, developed into accepted forensic tools, beginning in the late 1800s, forensic entomology was seldom practiced.

Yet the application of insect evidence to criminal investigations is not a new idea. A form of forensic entomology was practiced at least as early as the thirteenth century. In 1235 a Chinese "death investigator" named Sung Ts'u wrote a book entitled *The Washing Away of Wrongs,* which was translated into English by B. E. McKnight in 1981. Sung tells of a murder in a Chinese village in which the victim was repeatedly slashed. The local magistrate thought the wounds might have been inflicted by a sickle. Repeated questioning of witnesses and other avenues of investigation proved fruitless. Finally, the magistrate ordered all the village men to assemble, each with his own sickle. In the hot summer sun, flies were attracted to one sickle, because of the residue of blood and small tissue fragments still clinging to the blade and handle. Confronted with this evidence, the owner of the sickle confessed to the crime. The magistrate's action demonstrates considerable knowledge of the activity patterns of the flies, which were certainly blow flies. Indeed, in other parts of the book, Sung talks about the blow flies' activities in natural body openings and wounds, including an explanation of the relationship between maggots and adult flies, and discusses the timing of their invasion of a corpse.

Not until several centuries later, in 1668, was the link between fly eggs and maggots discovered in the West. Before then, people did not realize that maggots hatched from the eggs flies laid on exposed meat or decomposing bodies. Francesco Redi's studies of meat that was exposed to flies and meat that was protected from flies resulted in a major discovery. His observations of fly infestations on the exposed meat demonstrated the link between the flies' egg laying and the maggots, and disproved the concept of spontaneous generation. Before Redi, maggots were believed to arise spontaneously from rotten meat, not emerge from fly eggs. If my telephone log at the University of Hawaii at Manoa is any indication, quite a few people still think that maggots are worms with no connection to flies; one man even told me that maggots normally live "inside people" and only come out after we die.

Unfortunately, Redi's discovery did not lead immediately to the use of entomological evidence in death investigations. The first record of the use of insects in a forensic investigation in the West dates from 1855. During a remodeling of a house outside Paris, the mummified body of an infant was discovered behind a mantelpiece. Suspicion soon centered on the young couple then occupying the house. An autopsy was performed on the infant by Dr. Bergeret d'Arbois of nearby Jura, Switzerland, who concluded that the child had died in 1848. He noted evidence that a flesh fly, *Sarcophaga canaria,* had exploited the body during the first year (1848) and that mites had laid their eggs on the dried corpse the following year (1849). Bergeret's analysis of the insect evidence demonstrated to the satisfaction of the police that the death had occurred much earlier than 1855 and that the logical suspects were the occupants of the house in 1848, and they were subsequently arrested and convicted of the murder. The methods employed by Bergeret d'Arbois in his analysis are essentially the same as those forensic entomologists use today in estimating the time since death. He recognized and drew conclusions from the predictable pattern of succession of different insect species onto a corpse, and saw the significance of the duration of the life cycles of different carrion-frequenting insects. But although the assumptions underlying his analysis were correct, he appears to have misinterpreted some of the life cycles of the insects involved, and his conclusions would probably not have passed muster in a modern courtroom.

FORENSIC ENTOMOLOGY SHOULD logically have continued to develop steadily, with new discoveries arising from the findings of previous research. But progress was uneven and sporadic. Research was usually conducted in response to a murder and often ceased once the case was solved. There were some exceptions, particularly the work of J. P. Megnin in France. During the late 1800s, he published a series of papers on medicocriminal entomology that alerted both doctors and lawyers to the usefulness of entomological evidence. Possibly the most significant of these papers was *La faune des cadavres: Application l'entomologie a la medicine legale,* which was published in 1894. The central thrust of Megnin's work was that the postmortem interval can be determined by analyzing the various species of arthropods—invertebrates with a segmented body and paired, jointed legs, such as the insects, the mites, and the spiders—that are present on a decomposing body, without regard to their age. Today's forensic entomologists recognize this principle, but usually also take into account the age of each species in determining the postmortem interval.

In the mid-1930s entomological evidence came to the fore in a particularly brutal murder case, recently chronicled in *New Scientist* by Zak Erzinçlioglu. On September 29, 1935, a woman spotted a severed human arm while looking over a bridge spanning a small stream in Scotland. Ultimately, over 70 pieces of two badly decomposed corpses were recovered from the area, since known as the Devil's Beef Tub. The reassembled parts were eventually identified as the remains of Isabella Ruxton, the wife of a local physician named Buck Ruxton, and her personal maid, Mary Rogerson. Mrs. Ruxton had last been seen alive on September 14. Among the varied pieces of evidence collected at the scene was a group of maggots feeding on the decomposing body parts. These maggots were sent to a laboratory at the University of Edinburgh, where A. G. Mearns identified them as maggots of a blow fly, *Calliphora vicina,* and estimated that they were between 12 and 14 days old when they were collected. Since these maggots had developed from eggs laid on the body parts by adult flies in the vicinity, the bodies could not have lain near the stream for less than 12 to 14 days. This was the minimum time between the deaths and the discovery of the dismembered bodies. Suspicion fell on Buck Ruxton for a number of reasons, and the entomological estimate of the time of death became highly significant. The two bodies had been dismembered so skillfully that authorities believed the perpetrator had some knowledge of the human anatomy. Dr. Ruxton had been seen with a cut finger on September 16. In addition, one of the severed heads was wrapped in clothing belonging to one of the Ruxton's children, and the cleaning woman reported blood stains and foul odors in the Ruxton residence on September 17. All the evidence pointed toward Ruxton as the murderer, and the insects had been instrumental in pinpointing the time of death. Although he never confessed, Buck Ruxton was convicted of both murders and hanged.

More recently, Pekka Nuorteva of Finland has made important contributions to forensic entomology. The majority of his cases were murders, but one case he related in 1977 dealt with a dispute over the cleaning of government offices. During the summer, an official in the Finnish government noticed a lot of large maggots under the carpet near his office door. The official called for the cleaning woman and asked how often she cleaned the carpet. She replied that the carpets were cleaned daily and that she had last cleaned his office the evening before. The official could not believe that maggots over 1 centimeter long had developed overnight, and the cleaning woman was dismissed for lying and not properly performing her duties. As a matter of curiosity, a veterinarian was asked to look at the maggots and the carpet. He did not believe that the maggots could have developed by feeding on the synthetic fibers of the carpet, so he collected some of the larvae and sent them to Nuorteva for his opinion. The larvae were post-feeding maggots of a species of blow fly, *Phaenicia sericata.* At this stage of its development, a blowfly maggot stops feeding and migrates away from the food source to pupate. Nuorteva concluded that these larvae had probably developed on the carcasses of mice in the building or on some improperly stored food materials, definitely not on a synthetic carpet. During the night, they had migrated away from their food source and onto the carpet. Given this information, the official rehired the cleaning woman, but there is no record of any apology.

IN THE UNITED STATES, Bernard Greenberg of the University of Illinois at Chicago is widely regarded as the father of forensic entomology. His original training was in acarology, the study of mites, at the University of Kansas. But after obtaining his master's degree, Greenberg shifted his attention to the blow flies that constitute the family Calliphoridae and he is now a world authority on this family. Over the years he has been an innovative researcher and mentor to a number of graduate students. His research on the biology and life cycles of the many species of blow flies provided a strong basis for forensic work, and for many years he was essentially the only entomologist in the United States who devoted a major part of his research effort to issues important in forensic entomology.

During the mid-1960s, a graduate student in North Carolina named Jerry Payne began to lay the groundwork for another modern approach: succession. Simply stated, succession is the idea that as each organism or group of organisms feeds on a body, it changes the body. This change in turn makes the body attractive to another group of organisms, which changes the body for the next group, and so on until the body has been reduced to a skeleton. This is a predictable process, with different groups of organisms occupying the decomposing body at different times. In his landmark paper, published in the journal *Ecology* in 1965, Jerry Payne detailed the changes that occurred during the decomposition of pig carcasses that were exposed to insects, compared to the changes in pig carcasses that were protected from insect activity. This work built on and refined studies conducted almost 70 years earlier in France by Megnin. Megnin recognized nine stages of decomposition, but Payne recognized only six, introducing the system currently used by most forensic entomologists. Payne further emphasized the great variety of organisms involved in the process, recording over 500 species. Payne also conducted other experiments, including studies of submerged carrion.

Although others had studied decomposition before Payne, his work has received the greatest attention, possibly because of the thoroughness of his research, but also in part because of the status and circulation of the journal *Ecology.* Other excellent studies of decomposition—such as work by H. B. Reed, Jr., on insects involved in the decomposition of dog carcasses and G. F. Bornemissza's work on the effects of guinea pig decomposition on soil-dwelling arthropods—have appeared in regional journals and not received the widespread attention they deserved.

A number of other excellent biological and succession studies were published from the 1950s on, but none of these was undertaken with its forensic potential in mind. These studies, frequently conducted by graduate students, usually focused on agricultural or public health problems: In the 1950s and 1960s, there was much concern about the possible aftermath of an atomic attack in a society with few provisions for dealing with the public health problems presented by the accumulation of millions of dead bodies.

While entomologists were largely ignoring forensics in their publications, physical anthropologists were concentrating on the changes that occur in a human body between death and skeletonization. They recognized the major roles played by insects in the decomposition processes. Initially, the subject was discussed only in notes and in observations of isolated cases involving insects. But with the establishment in 1981 of William Bass's Anthropological Research Facility at the University of Tennessee, controlled studies of human remains began to be conducted. Work by William Rodriguez and Bill Bass during the 1980s yielded significant new information about the colonization of exposed and buried bodies by insects.

IT WAS IN the early 1980s that I became involved in forensic entomology. Before then, I had been conducting research in acarology. I had completed my Ph.D. in entomology at the University of Hawaii at Manoa in 1977, and moved immediately from the university to the privately endowed Bernice Pauahi Bishop Museum some 7 miles across town. There I became the principal investigator for a study of the classification of the chiggers of Papua New Guinea funded by a grant from the National Institutes of Health. I was dismayed to discover that the NIH grant paid less than I had earned as a graduate student. To supplement my income, I took on a series of side jobs, ranging from occasional entomological consultations to providing security for wholesale gift and jewelry shows.

One of the benefits of the NIH grant was a stipend to attend the annual meetings of the Entomological Society of America. In December 1981 the meeting was held in San Diego at the Town and Country Hotel. Since I was on a limited budget, I stayed at another hotel eight freeway lanes and three fences away from the main meeting hotel. This meant either I had to walk 2 miles to get to the meetings without crossing the freeway through traffic or I had to dodge eight lanes of speeding cars to get to the other side. I walked the 2 miles. As a result, I went to the meetings early in the morning and did not return to my hotel until late in the evening, an arrangement that led to my attendance at several presentations I normally would have missed. Among these was an early-morning presentation by Lamar Meek of Louisiana State University.

Meek talked about the research he was conducting on decomposition, ending with a discussion of a murder investigation. He illustrated the first part of his talk with slides of decomposing pigs, then shifted to slides of human remains. As the slide show went on, I began to think that this field might hold some promise for me. I already had my Ph.D. in entomology, and I had had first-hand experience with dead bodies during my stint in the U.S. Army. I had been drafted a couple of months after earning my bachelor's degree in zoology, and I spent most of my 2 years in the military in the morgue at the U.S. Army Hospital at Fort Ord, California, but also had a couple of tours of duty with a chemical, biological, and radiological warfare unit at Edgewood Arsenal in Maryland. There I was a subject in tests of riot-control gasses, not the most enjoyable period of my life. This combination of activities left me with a better working knowledge of human anatomy than I had acquired in formal classes. By 1981, in short, I had some knowledge of insects and was fairly confident of my ability to deal with dead bodies. More important, forensic entomology appeared to be a more interesting line of work than giving farmers advice on how to keep spider mites from damaging tomato plants.

On my return to Honolulu from San Diego, I began to look for background material and names of other people working in the field. There was not much in the entomological literature specifically devoted to forensic entomology. I found most of the older references and a few more recent papers, but it quickly became apparent that if I really wanted to pursue this field, I would have to involve myself with people I had not dealt with on a cooperative basis before. Since my only prior experience with the police was being on the receiving end of traffic citations, I decided to start with the medical examiner.

The medical examiner in Honolulu at the time was Charles Odom. He had a particularly diligent secretary screening his calls, and for a couple of months whenever I called Odom was out of town. Eventually, I did manage to get in touch with him and persuade him to meet with me. We had lunch together at the Willows Restaurant near the University of Hawaii at Manoa. By the end of the meal we had decided that if we could discuss maggots and decomposing bodies while eating curry over rice, we could probably work together.

I soon discovered that agreeing to work together and actually beginning to work together are not the same. For one thing, despite the impression conveyed by such late-night television staples as *Hawaii Five-O* and *Magnum, P.I.*, the homicide rate in Hawaii is extremely low compared to that in the other 49 states. And on Oahu, where I do most of my forensic work, the bodies of murder victims are usually discovered very quickly because the island is relatively small and because the terrain and the hot and humid climate do not favor lengthy concealment of corpses. For these reasons relatively few death investigations in the city and county of Honolulu involve bodies that have been decomposing for more than a brief period of time. Consequently, at least at the beginning, I became involved in murder cases only if I contacted the medical examiner's office whenever I heard or read in the newspapers about a death involving a badly decomposed body.

Apart from this obstacle, I met surprising resistance from the Bishop Museum. The feeling was that I might involve the museum in situations that would generate unfavorable publicity and perhaps even lawsuits resulting from my testimony. I found both concerns hard to understand, particularly the bad publicity: it's difficult to see how helping solve a murder can be considered "bad." While I was attempting to overcome this resistance, a solution arose in the form of an opening in the Department of Entomology at the University of Hawaii at Manoa.

The College of Tropical Agriculture and Human Resources seemed to be the perfect place for forensic study. The position required me to teach courses in acarology and classes in medical and veterinary entomology (about insects and other arthropods as the causes or vectors of human and animal diseases), and help ensure the well-being of agriculture in the state of Hawaii. I was well qualified for the teaching requirements, but the university's definition of agriculture at that time seemed restricted to the activities of farmers growing plants for food. So I redefined both "agriculture" and "human resources" in my own mind and accepted the position. There were several advantages to my new position, not the least of which was a sense of financial and professional stability—grant funds, and with them the job, have a way of expiring on very short notice. Moreover, the university provided me with a base of operations that was more acceptable both to the police and to the medical examiner than a privately funded museum. An unforeseen bonus was the potential for graduate students to assist in research projects. At any university and in any department there are always some students willing to explore new and unusual areas. Over the years, I've managed to attract more than my share of these students to my projects and their work has always been productive.

At about the same time, several other entomologists and parasitologists began to explore forensic entomology. Some, such as Paul Catts, had been consulted previously on criminal cases. Others, such as Wayne Lord, had spent more time investigating the ecological aspects of decomposition. I myself was still exploring the field. Of the group, only Bernard Greenberg was involved in forensic entomology on any kind of continuing basis.

Over the next couple of years, we began to gravitate toward each other during the annual meetings of the Entomological Society of America, assembling for the first time in 1984 at the annual meeting in San Antonio. Lamar Meek had organized a symposium on forensic entomology, and after the morning session we decided to lunch together. This was the beginning of a tradition. Initially, we got together for lunch, but we soon shifted to breakfast meetings. Sitting around the table, we created an informal group called CAFE. Theoretically, this stood for Council of American Forensic Entomologists, but in practice it meant that we met in various cafés. The group had no official organization although Paul Catts usually made any reservations for tables. The only real criteria for membership were being able to get up for breakfast and being able to look at pictures of decomposing bodies while eating. Early on, Paul Catts began to refer to the group as the Dirty Dozen and the name stuck.

This group has grown over the years and now numbers about 15 people in the United States and Canada who are routinely involved in forensic entomology. This informal association has provided a forum for the exchange of ideas and moral support in our studies. Many of the techniques I talk about later in the book were developed as a result of the interactions between members of this group. After 12 years, in 1996, we finally formed an official organization, the American Board of Forensic Entomology.

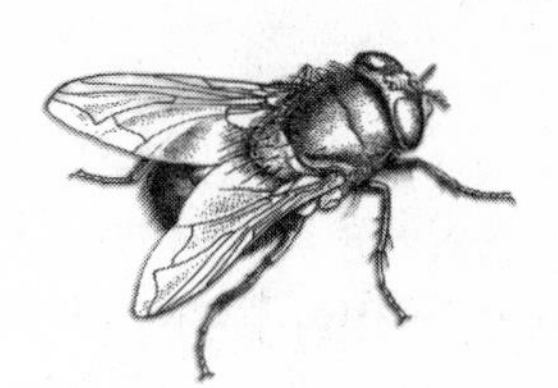

2

The Bugs on the Body

A decomposing body is in some ways like a barren volcanic island that has recently emerged from the ocean. The island is a resource, isolated from similar areas, waiting to be colonized by plants and animals. The first plants establish a beachhead and begin to change the island, making it habitable for later arrivals. Similarly, a dead body is a resource that is usually isolated from other dead bodies by dissimilar patches of habitat, such as fields, ponds, and woodlots. Unlike an island, though, a decomposing body is a temporary microhabitat, a rapidly changing and disappearing food source for a wide variety of different organisms, ranging from microscopic bacteria to fungi to large vertebrate scavengers, such as feral dogs and cats. Of course some organisms, such as the bacteria present in the intestines, exist in the living body, but the animals that invade and consume a dead body form a distinct group. Like the island, the dead body has definite boundaries, and all the changes that occur take place within or very close to the body.

The majority of the carrion animals are arthropods, and among those arthropods found on a decomposing body, the insects are the predominant group in terms of numbers of individuals present, biomass (total weight of the individuals), and diversity (number of different species). On average, about 85 percent of the species reported in decomposition studies are insects.

Insects and other arthropods can be associated with a corpse in many ways, but forensic entomologists agree on four main types of direct relationships and categorize the carrion species accordingly. The first category consists of the necrophagous species, those that feed directly on the corpse, primarily flies (Diptera) and beetles (Coleoptera). Flies, especially the blow flies and flesh flies, depend on decomposing matter for food. These flies are aggressive in their search for human and animal remains and frequently arrive mere minutes after death. During the first 2 weeks of decomposition, the blow flies and flesh flies are usually the most precise indicators of the postmortem interval. Many beetles are no less dependent on decomposing matter, but they usually arrive later in the decomposition process, after the body has begun to dry.

As the populations of the species that feed directly on the corpse increase, they attract another group of arthropods, the predators and parasites of the necrophagous species. These animals are attracted not to the dead body, but to the other insects already feeding on it. Among the first of this group to arrive are the burying beetles (family Silphidae), the rove beetles (family Staphylinidae), and the hister beetles (family Histeridae). They all prey on the eggs and maggots of flies feeding on the dead body. Some flies are also predatory during their larval stage, such as the soldier flies in the family Stratiomyidae.

Some species manage to act both as necrophages and as predators. One is the blow fly *Chrysomya rufifacies,* which you met earlier. It is well distributed throughout Asia and the Pacific islands, and has recently been introduced into Central and South

America and the southern United States. It is one of the most common species of blow flies found on dead bodies in Hawaii, often arriving within 10 minutes of death. Another blow fly in the same genus, *Chrysomya megacephala,* also arrives almost immediately after death. Females of both species light on the body, feed on any blood or fluids that are available, and then start laying their eggs in and around the natural body cavities. In the field, I have observed that if *Chrysomya rufifacies* arrives before *Chrysomya megacephala,* it delays egg laying until after *Chrysomya megacephala* arrives. Politeness is not a consideration here. *Chrysomya megacephala* can feed *only* on the decomposing body, but *Chrysomya rufifacies,* although it prefers to feed on the decomposing body, will, if that food source is exhausted, change life styles and become a predator. And one of its favorite prey species appears to be *Chrysomya megacephala.* As decomposition progresses under typical conditions in Hawaii, it is not unusual to find only *Chrysomya rufifacies* maggots present on the body.

The parasites associated with the necrophagous species are primarily representatives of the Hymenoptera, the order of insects that includes the ants, bees, and wasps. A large number of very small wasps (often less than 1 millimeter long) are parasitic on the maggots and pupae of flies. These tiny wasps lay their eggs either inside or on the outside of the maggot or pupa. The eggs

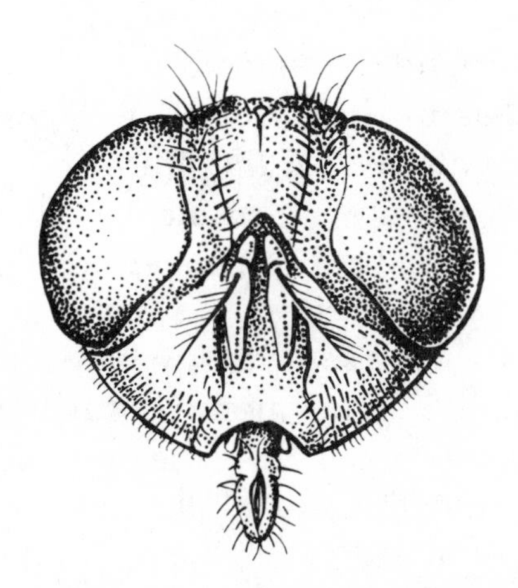

Detail of the head of the female blow fly *Chrysomya megacephela*

hatch and the wasp larvae feed on the developing fly. The eggs of these parasites frequently produce more than one adult wasp (the number depends on the species of wasp), and up to several hundred wasps may emerge from a single parasitized maggot or pupa. Of importance to forensic entomologists, these wasps frequently specialize on a particular type of fly, and therefore they can provide clues as to what flies were actually on a body even after the flies have completed their development and departed.

The third group of animals associated with a dead body are the species, such as wasps, ants, and some beetles, that feed on both the body and the other arthropods. Unlike *Chrysomya rufifacies,* these species are omnivorous and feed on both types of food continuously. During decomposition studies conducted in 1985 inside Diamond Head Crater in Hawaii, I observed some very aggressive ants, *Solenopsis geminata,* removing maggots from my test carcass in such great numbers that they actually slowed the rate of decomposition. Adult wasps are especially active around a decomposing body during the early stages of decomposition. They feed on adult flies, often capturing them while still in the air, and also on fluids from the body. In some of the studies I have conducted on the island of Hawaii, wasps of a species that evolved in Hawaii, *Ectimneus polynesialis,* were so efficient at capturing flies that their activities delayed the onset of decomposition for over a day.

Among the arthropods I frequently encounter during investigations are those that use the corpse as an extension of their normal habitat, representatives of the fourth category of animals associated with a dead body. These include hunting spiders that congregate on and around a corpse to prey on the various insects present. Occasionally, a spider even uses part of the corpse as an anchor for its webs. There's something a bit surreal in the sight of dew drops glistening in the morning sun on a spider web attached to a decomposing arm. As the body decomposes, fluid by-products seep into the soil under the body. This process begins early in decomposition and continues until the body has dried completely. These fluids provide nutrients for a large number of soil-dwelling organisms, and there is always a distinct population of decomposition-associated organisms present in the soil under the body, often persisting for several years after the death. Of major significance among these are the various species of soil-dwelling mites (Acari), the springtails (Collembola), and the roundworms, or nematodes.

As if to confuse the picture, there may also be arthropods on the corpse that have nothing to do with decomposition. Some may have fallen onto the body from surrounding vegetation, especially if the murderer disturbed undergrowth in an effort to conceal the body. And some may simply have landed on the corpse because it was a handy landing spot when they stopped flying.

IT IS THE job of the forensic entomologist to interpret the varied interactions between arthropods and the corpse during decomposition, eliminating from consideration organisms that are present by accident, and to provide law enforcement officials with information that they can use to apprehend and convict murderers. Usually, the most important and fundamental contribution of the entomologist is to determine the time since death, the postmortem interval—as you saw in the case of the body discovered by the three fishermen. But forensic entomologists and the insects they study may also provide other valuable information that will help solve the crime. Insects can provide valuable clues about the movement of a body following death. Insects are found in virtually every habitable part of the earth. But not all insects occur in all types of habitats. Some are quite specific to a given type of climate, vegetation, elevation, or time of year. In temperate regions finding an insect that is typically active during the fall on a body discovered in the spring indicates that the death occurred during the fall. If, as occasionally occurs in Hawaii, an insect specific to an urban habitat is found infesting a body discovered in an agricultural or rural area, investigators can be fairly certain that the crime was not committed at the scene of discovery. Instead, the murder probably took place in an urban area, and the body was later dumped where it was eventually discovered.

One such case occurred on Oahu, where the decomposing body of a woman was found in a sugar cane field. The majority of insects recovered were species found all over the island, but there were maggots of one species of fly that in Hawaii is typically found on a decomposing body only in urban dwellings, *Synthesiomyia nudiseta.* These were also the most mature maggots on the body. I concluded from this evidence that the woman had been killed in an urban area, and the body had been exposed to insect activity there for some time before being deposited in the sugar cane field. The developmental stages of the majority of the maggots present on the body suggested that it had probably been in the cane field where it was discovered for about 3 days, but the developmental stages of the *Synthesiomyia nudiseta* maggots indicated that the postmortem interval was in the range of 5 days. When the victim was identified, it was determined that she had been killed in an apartment in Honolulu when a drug deal went bad. The murder was not planned, and the murderers kept the body in the apartment for a couple of days while deciding how to dispose of it. In this case, the insects provided an estimate of the time of death as well as an indication of the location.

The invasion of a dead body by insects follows a definite pattern if the body is intact, without open wounds or external bleeding. First to be invaded are the natural body openings: the eyes, mouth, nose, and ears, followed by the anus and genitals if they are exposed. Blood or wounds provide additional points of entry and insect activity. Wounds inflicted before death (antemortem injuries) or at the time of death (perimortem injuries) are more attractive to insects than those inflicted after death (postmortem injuries) because they bleed, often profusely. Wounds inflicted after death, when the heart is no longer pumping, produce little if any blood, and are not as attractive to insects. Insect activity may alter the characteristics of any kind of wound, but insect activity during the early stages of decomposition associated with areas other than natural body openings should alert investigators to the possibility of antemortem or perimortem wounds.

Consider a case that occurred in Tennessee some years ago. The body of a young woman was discovered in such an advanced state of decomposition that the local coroner could not determine the cause and manner of death. Investigators did, however, note that there were peculiar patterns of insect invasion in the chest and in the palms of the hands. They referred the case to a physical anthropologist and an entomologist. Both experts agreed that these patterns were unusual and warranted further examination of the body. The body was exhumed and a reexamination of the skeleton revealed evidence of cuts to the ribs and the bones of the hands consistent with stab wounds to the chest and defense wounds to the palms of the hands as the victim tried to fight off her attacker. Without the entomological evidence, the cause and manner of death in this case would probably never have been discovered.

Entomological evidence may also place a suspect at the scene of a crime. An unusual case from Texas, described to me by a retired FBI agent, involved the body of a woman found with the mangled remains of a grasshopper in her clothing. At first, nobody paid much attention to the grasshopper, although its parts were collected and preserved as evidence. The police identified several suspects and brought them in for questioning. At the time, 1985, male fashion was making another of its major statements by reintroducing cuffs on men's pants. During a search of the suspects, the left hind leg of a grasshopper was discovered in the cuff of one suspect's pants. This was the only part of the grasshopper that had not been recovered from the body, and the fracture marks matched perfectly. Despite the defense attorney's assertion that "grasshoppers always break their legs like that," the suspect was convicted of murder.

Expert knowledge of arthropod activity played a major role in placing a suspect at a crime scene in a rape and murder case in

California. In August 1982, a search and rescue team discovered the nude body of a 24-year-old woman beside a dirt road in a rural area outside of Thousand Oaks. The body was lying under a large eucalyptus tree and on a slight incline, close to a level field of wild oats. The corpse was partially obscured by broken branches and other underbrush, apparently placed there by the murderer. The woman's blouse was tied around her neck. Since the body was discovered at night, it was left undisturbed until daylight the following morning, when an efficient search of the crime scene could be conducted. Although the body was not moved, a preliminary survey of the area was conducted by a sergeant from the Ventura County Sheriff's Office Homicide Team from 10:00 P.M. that night until 2:00 A.M. the next morning. Later that morning the sergeant noticed a number of red, inflamed bites on his ankles, waist, and buttocks. They reminded him of chigger bites he had received years before while he was on military duty in Kentucky. He later discovered that 20 of the 23 members of the search and rescue team had similar bites. Chiggers that attack humans are not common in southern California, so these bites were unusual. The autopsy revealed that the woman had been strangled and that sperm were present, confirming the possibility of rape as a motive for murder. But no chigger bites were observed on the body during the autopsy.

Then, while viewing photographs taken during strip searches of suspects in the case, the sergeant noticed bites similar to his own on the lower legs, waist, and buttocks of one suspect. He thought there might be a connection between his bites, the bites on the members of the search and rescue team, the bites on the suspect, and the crime scene. To check, he contacted the School of Public Health at the University of California at Los Angeles, which referred him to the Chigger Research Laboratory at California State University, Long Beach. There Dr. James Webb agreed to investigate. With several lab workers Dr. Webb went to the crime scene to collect chiggers.

Chiggers are the larvae of mites in the family Trombiculidae. These mites have an unusual life cycle. The chiggers are external parasites that suck tissue juices from their hosts, such vertebrates as lizards, rodents, birds, and people. The chigger attaches to a host and takes a single meal before detaching and dropping to the soil to complete its development. In humans, chigger bites cause a localized allergic reaction, resulting in red welts. These bites tend to occur on humans where clothing is in close contact with the skin and, I'm told, "itch like Hell." (Oddly, though I've worked on chiggers for over 25 years, I've yet to be bitten.) In the later stages of their life cycle (nymphs and adults), these mites are predators, feeding on small soil-dwelling arthropods and their eggs. These predatory stages have very specific living requirements with respect to the soil pH, relative humidity, and temperatures, and thus have limited distributions. Indeed, some are so highly adapted to a habitat under scrub brush that they cannot survive in the grasses as little as 15 feet away. The next generation of chiggers produced will remain in this habitat while waiting for a host. Thus the presence of chigger bites on a suspect may well indicate he has been at a very specific location.

In their sampling, Webb and his team set traps for small mammals around the crime scene, collected lizards, set black plates (devices specifically designed to catch chiggers), and, unintentionally, acted as chigger bait themselves. Six of the seven lizards collected were infested by a species of chigger, *Eutrombicula belkini,* which is known to attack humans in California. Rodent traps yielded desert wood rats, deer mice, and a pocket mouse. Only the pocket mouse and one of the desert wood rats were parasitized by *Eutrombicula belkini.* The same species of chigger was also collected from the black plates. During the sampling, the team members observed chiggers moving on their boots, and one unfortunate team member proved to be an excellent attractant for chiggers, being bitten 43 times. These chiggers were not collected in significant numbers from other areas surrounding the crime scene.

During questioning, the suspect said that he had been bitten by fleas in his sister's house in Thousand Oaks. The team went to her house and did the same sampling and trapping they had done at the crime scene. The house was in a suburb of Thousand Oaks; wild grasses grew along the garage, domesticated grasses in the lawn, and oleander bushes along the side of the lot. The team set traps along the garage and fences, and sampled the lawn with black plate traps. No small mammals were captured, and the only lizard collected was an alligator lizard, which was free of any external parasites. The black plates did not yield any chiggers. In fact, the team could not find any chiggers, fleas, or other arthropods that could account for the bites on the suspect.

During a preliminary hearing in October 1982, Webb presented the entomological findings. The suspect had a prior history of sexual assault, but there was little of physical evidence aside from that provided by the chiggers to link him to the crime. The entomological evidence was considered sufficient, and the suspect was indicted and tried for rape and murder in February 1983. Citing the evidence from the police investigation, the autopsy report, and the entomological investigation, the prosecution contended that the suspect had raped and killed the woman at the scene. While attempting to conceal the body with the grasses, he had been bitten by the chiggers. The defendant admitted that he had been with the woman early on the evening of August 3, the last time she had been seen alive, but claimed he had left her alive, and said he had never been at the site where the body was found. The defense attorney suggested that the bites could have been inflicted by other insects, possibly fleas. The prosecution maintained, correctly, that the distribution of the bites on the defendant's body was typical of chigger bites and completely atypical of flea bites. This, combined with the very limited distribution of the chigger *Eutrombicula belkini* in Ventura County, virtually eliminated the possibility that the defendant could have been bitten elsewhere. The jury convicted him of rape and murder and he was sentenced to life in prison without parole.

3

THE PIGS' TALE

The insects involved in the decomposition of a body in the Hawaiian Islands are not usually the same as those infesting a corpse in Illinois, southeast Asia, or Europe. A few species may be common to such different locations and the groups of insects represented may be similar, but most species and life cycles are different. Outside the tropics, the insects on a corpse also vary according to the season of the year. In the summer blow flies in the genus *Phaenicia* are often the primary invaders of a corpse, while in the fall they may be almost completely absent, replaced by blow flies in another genus, such as *Calliphora.* Still other variations may result from the surroundings. The insects present in the inner cities are frequently quite different from those encountered in the country, and the use of pesticides in agricultural areas may further complicate the picture.

To determine the differences in decomposition in different habitats in Hawaii, I have conducted controlled decomposition studies on animals from which I could extrapolate when I was estimating the postmortem interval in homicide cases. In designing such studies several factors must be taken into account. First is the choice of the animal. For the results of the study to be usable, it must be an animal whose decomposition closely approximates the stages of human decomposition. And it must be readily available in large numbers so that studies can be replicated. Also the animal must be relatively cheap to obtain because funds are usually limited. Finally, it must be an animal whose carcass will not unduly upset any members of the community who may encounter an experiment in progress while on an afternoon walk or weekend hike.

Before I chose a test animal I read about decomposition studies conducted by others. Such studies have been carried out in various sites around the world with a wide variety of animals, ranging from toads and lizards in B. W. Cornaby's studies in Costa Rica to elephants in Malcolm Coe's study in Africa. Research has also been done on mice, birds, guinea pigs, rabbits, cats, dogs, pigs, and even human cadavers. Needless to say, the results from these varied carcasses have been quite different. Direct comparisons between these animals and humans are relatively scarce, so I compared results from individual studies conducted in similar areas and times, using data from the various animal studies and from homicide cases in which the time of death was known. On the basis of these data, the animal that seems to most closely approximate patterns of adult human decomposition is a domestic pig weighing about 50 pounds. This is the animal I use and it is now the test animal of choice for most decomposition studies in the United States.

In addition to selecting an appropriate animal, I wanted to choose environments similar to those where corpses have been discovered or are likely to be discovered. After all, data from a desert cannot properly be applied to a corpse found in a city or a pine forest without appropriate adjustments. I have been made aware of at least one instance where my data from Hawaii were applied to a case in Florida without adequate consideration of differences in either geography or environmental conditions. My data, derived from a decomposition study conducted in a lush rain forest on the island of Oahu, were applied to a corpse found on a nearly barren Florida sandbar. Aside from the obvious problems with that comparison, the fact that a hurricane had moved through the area during the period in question was not considered significant by the person providing the estimate of the postmortem interval. Needless to say, the estimated time since death bore no resemblance to the time that had really elapsed.

To find the most useful areas for decomposition studies, I examined records from the medical examiner and police on Oahu to determine where on the island corpses had most often been discovered. I then chose study sites in similar areas. Interestingly, the number of different habitats used historically was relatively small. In most cases, though the murderers appeared to have given considerable thought and planning to the events leading up to and including the murder, they apparently gave relatively little thought to the disposal of the body. Most frequently, bodies are either left exposed along roadsides, especially the sides of access roads through sugar cane or pineapple fields, or dumped in places close to trails or roads, especially the edges of ravines in somewhat deserted areas or parks. And sometimes, of course, bodies are hidden in buildings. I identified study sites that approximated each of these areas and that were relatively secure from disturbances, and with the help of my graduate students, began to conduct studies. To date, we have been successful in conducting studies at all these types of sites except the inside of a building.

In selecting sites, security has always been a major concern. So far, I have had excellent cooperation from the military and the State of Hawaii in obtaining study sites. I have always been careful to avoid areas where the general public might come into contact with our carcasses. I don't want to cause discomfort to any passers-by, and neither do I want to expose our sites to vandalism, which is sometimes a problem. In one homicide case I investigated, a body managed to remain in a fairly large metal tool box alongside a well-traveled road on the island of Oahu for 18 months without being disturbed; but a hygrothermograph, an instrument that records temperature and relative humidity, later placed at the crime scene in a much smaller box, disappeared in less than 24 hours. Similarly, near the university campus, my exclosure cages used to protect test animals from scavengers were removed in less than 6 hours, but the pigs were left behind. It seems that the wire mesh I had used to make the cages was the perfect size for fish traps.

Once the site has been selected and security arrangements made, the study can begin. At each study site, I use three 50-pound domestic pigs. Since I was attempting to duplicate a homicide, in one of my first studies I wanted to shoot each pig through the head with a 38-caliber pistol. This plan naturally required permission from the University of Hawaii's Institutional Animal Care and Use Committee. Needless to say, this was not on their list of approved methods of euthanasia for laboratory animals, and I was required to explain my logic during one of the committee's meetings. As I explained how I would be shooting the pigs and my reasons for needing them to be killed that way, I noticed several of the committee members moving their chairs away from the table and me. Clearly they were uncomfortable with my plans. Their major concern was the welfare of the pigs, and I cannot fault that in any way. My ultimate objective is to try to prevent murders, not to torture pigs. One member suggested that the pigs be given tranquilizers before being shot. I couldn't agree to that request because any drugs administered to the test animals might affect the insects feeding on the carcass. Another concern was that I might miss and have to shoot several times. In the end, it was decided that if I could find someone else—a police officer, a park ranger, or a hunter—to shoot the pig, I would be allowed to proceed. I did find a police officer willing to assist, so I was able to conduct the study.

In subsequent studies, I have obtained pigs from commercial pig farms on the island of Oahu. These farms are certified by the U.S. Department of Agriculture and the pigs are inspected by USDA inspectors following death. Initially, I thought I had overcome the earlier problems but I was being foolishly optimistic. This time, since the pigs would already be dead when I got them, the university's Institutional Animal Care and Use Committee tended to view the pigs as large pork chops purchased from the local supermarket and thus ethically acceptable. But since my studies were part of a project within the University of Hawaii at Manoa's College of Tropical Agriculture and Human Resources, a project outline was sent to the U.S. Department of Agriculture for review by its Cooperative State Research Education and Extension Service. This body examined the project proposal and then contacted the director of the college's Hawaii Institute of Tropical Agriculture and Human Resources with its major concern: Could I certify that the pigs were being killed humanely? This circuitous form of logic was mind-boggling. Neither the Institutional Animal Case and Use Committee nor I thought that we should be the ones assuring the USDA that a facility approved and inspected by the USDA was in compliance with USDA regulations. Ultimately, the large pork chop view prevailed, and I was allowed to begin the studies.

BEFORE STARTING A decomposition study, I place the data-recording equipment needed near each exposure area. At the center of each site, I install equipment to record the environmental conditions during the study: a hygrothermograph to record ambient air temperatures and relative humidity; a rain gauge to record the daily rainfall; and a high-low thermometer to record daily maximum and minimum temperatures. I also take a representative soil sample that I will process to extract soil-dwelling species. These arthropods will serve as my baseline for comparison with later samples to demonstrate changes in the soil fauna during decomposition.

I place the three pigs at least 50 meters apart, because I have learned from previous studies that shorter distances do not prevent abnormal attraction patterns among the insects. If the pigs are placed closer together, one pig will usually have significantly greater insect activity than the other two, or two pigs will have more activity than the third. I have seen similar patterns of skewed invasion in cases where several bodies were discovered in close proximity. In these instances, one body will almost always have greater numbers of insects than the others.

The pigs are killed at 6:00 A.M. on the morning of day 1 of the study and transported to the study site inside two plastic bags to prevent insects from getting onto the bodies before the pigs are placed at the site. Since the island of Oahu is fairly small, no more than 15 to 30 minutes pass between death and exposure of the bodies to insects.

Each of the three pigs serves a different purpose during the study. One of the pigs is placed directly on the ground and thermocouple probes are inserted into its anus, abdominal cavity, and head. This pig is left untouched for the duration of the study. I take pictures of the pig each time I visit the site to record physical changes as decomposition progresses. During the decomposition process, there are significant changes in the internal temperature of the pig due to the activities of bacteria inside the body and the metabolic processes of the insects feeding on the body. I document these changes by recording at each visit the temperatures registered by the probes inserted into the pig. And I also record the soil temperatures near the body.

The second pig is used to determine the rate of removal of tissues, or biomass, during the decomposition process. This removal rate is most conveniently measured by changes in the weight of the pig over time, and, for comparison purposes, I express this loss as the percent of weight remaining. This pig is placed on welded wire mesh, reinforced by a frame of wooden dowels, that will support the pig during the weighing. This arrangement allows the pig to remain in contact with the soil. I use a scale suspended from a wooden tripod placed above the pig to weigh it every time I visit the site. Because changes occur in the soil under the body during decomposition, it is essential to return the pig to the same position on the ground after weighing. If this is not done, the pattern of insect activity under the body will be disrupted and the data will not accurately reflect what happens during decomposition. I make sure that the pig is returned to the same position by using guide stakes that I drive into the ground at each corner of the frame.

The third pig is also placed on wire mesh. This pig is used for sampling the insects and other arthropods. As with the second pig, the carcass must be returned to the original position after sampling to ensure that the data are accurate. To prevent vertebrate scavengers from disrupting the study by removing flesh from the pigs, I put each of the pigs under a welded wire mesh exclosure cage that is open at the bottom. In mainland areas, a number of different vertebrates will feed on a decomposing body. In Hawaii, there are a limited number of vertebrate scavengers, mostly feral cats and dogs, rodents, and mongooses. But although they are few, they are voracious, particularly the mongooses. These become quite bold and territorial as the study progresses and frequently come up to the carcasses while I am working at the site. It is disconcerting to be collecting maggots from one end of a pig and look up to find a mongoose eating at the other end.

For the first 14 days of each study, I visit the site at least twice a day. During all studies, one of these visits will be at 1 hour past the solar zenith. Since Hawaii is in the tropics there is very little change in day length over the year, and this time is always at approximately 1:00 P.M. The time of the other visits depends to some extent on the location of the site. Whenever possible, I try to visit two other times: once in the early morning and a third time in the late afternoon. After 14 days pass, I visit the site once a day for the next 21 days at 1:00 P.M. After that, depending on the rate at which changes appear to be occurring, I visit less frequently, although still at the same time of day. Each day at the 1:00 P.M. visit, I record the maximum and minimum temperatures and the daily rainfall.

At each visit, I first spend some time simply quietly observing what is happening to the carcasses. It's important not to rush up to the carcass and startle all the mobile species into flight before their presence can be recorded. I then photograph the pig placed directly on the ground and record the temperature readings from the probes inserted into the body, the ambient air temperature, and the soil temperature next to the body. For the second pig, I observe any insect activity, apparent, take photographs, and record the weight. The third pig serves as the major collection site, and I take samples of all the arthropods I can detect. Collections made from each site must be labeled in the field and kept separately, and it's imperative to segregate the predatory species from the necrophagous species. If you do not do this, you run the risk of having only a very fat and satisfied predator in the vial by the time you return to the laboratory to process specimens. Every third day, I take a soil sample from the area under the pig, which will later be processed in the laboratory to extract the arthropods.

Back in the laboratory, the work of preserving the specimens begins. Adult insects are either preserved in ethyl alcohol or dried and pinned, depending on the type of insect. Soft-bodied species are placed in a 70 percent ethyl alcohol solution, while hard-bodied insects are killed, dried, and mounted on insect pins. Each lot of insects must be labeled with the time and date of collection, their location on the body, and details of the site. Without this information attached, the specimens are useless for purposes of analysis.

The immature specimens are treated somewhat differently. Since it's often impossible to identify immature insects accurately to the species level because they look so much alike, I split the collections of immatures into two sublots. Immatures are usually soft bodied and easily injured, so the specimens must be handled gently. The first sublot goes into a rearing container with a food supply, usually beef liver, and the container goes into an environmental chamber in the laboratory. The chamber resembles a refrigerator, but can be programed to maintain constant temperatures from –30 to +70°C. The periods of light and dark within the chamber can also be programed. My chamber is normally set for 26°C, or 79°F, with 12 hours of light and 12 hours of darkness. Specimens are reared to the adult stage and correlated with the immature specimens in the second sublot.

Specimens in the second sublot of each collection represent the units for the "biological clock" I must interpret to determine the postmortem interval. This clock is started with the invasion of the insects and is stopped by the collection and preservation of insects from the corpse. Each stage of development of each insect or other arthropod collected from the body represents a distinct interval of time on that clock—hours, days, months, or possibly years. To determine the interval represented, each sample must be fixed in time and development by killing and preserving the specimens.

Killing and preserving immature insects or larvae is not always easy. Even though most larvae, such as maggots and caterpillars, are soft bodied, they have evolved effective ways to withstand the extremes and hazards of their environment. One of the major environmental problems faced by an immature insect is loss of water. The larva solves this problem by producing a waterproof layer on the outside of its external skeleton, or cuticle. This waterproofing helps the larva, but prevents preservation fluids from entering its body. If these fluids do not penetrate the cuticle, the larva begins to rot from the inside and eventually shrivels up and turns black. Specimens in this condition are useless for forensic purposes. To overcome this problem entomologists use a fixative, a fluid that breaks down the outer protective layer and allows the preservative to reach the inner tissues of the insect. The fixative also expands the larva, giving a more accurate picture of its actual degree of development by making details such as the setae (bristles) and spiracles more clearly visible. Over the years entomologists have used a number of different fixatives, many of which have recently been shown to be carcinogenic and are therefore no longer used. Today the fixative of choice is a mixture of kerosene, acetic acid, and ethyl alcohol, known as KAA. The immature insects are put into this solution and kept there for a few minutes to several hours, depending on their size. After this, they are transferred to 70 to 80 percent ethyl alcohol for final preservation and storage.

If for some reason my return to the laboratory is delayed and there is no fixative at hand, hot water will do in a pinch. The hot water needed is about the same temperature as hot coffee or tea, approximately 77°C, or 170°F. The easiest way to get it is to drive to the nearest fast food restaurant and get a cup of tea to go. Usually I am given a cup of hot water and a teabag. I discard the teabag and put the larvae into the hot water for a couple of minutes and then transfer them to 70 percent ethyl alcohol.

First instar larva

Posterior spiracles

Second instar larva

Posterior spiracles

Third instar larva

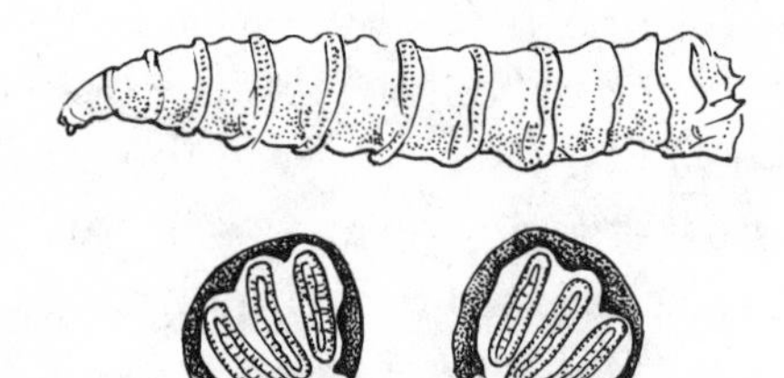

Posterior spiracles

Detail of *Chrysomya megacephala* instars showing spiracles

To collect samples of soil-dwelling arthropods, I use Berlese funnels. These funnels are named for the Italian acarologist Antonio Berlese, who designed the first ones in the 1800s. Made of metal or plastic, each funnel has a fine wire mesh across the top and a jar of alcohol or some other preservative fluid at the bottom. I put the soil sample on the wire mesh and suspend a light above it. Since soil-dwelling arthropods prefer cool, moist, and dark conditions, they migrate through the sample, away from the light and source of heat, until they finally fall through the mesh and into the jar of preservative. Soil samples processed in the laboratory in Berlese funnels yield quite small arthropods. Because of their tiny size, I usually mount these specimens on slides and identify them with the help of a compound microscope.

After preservation comes the major, and frequently boring, task of identifying the specimens. This is a matter of sitting for long hours over a dissection or compound microscope looking at various structures and counting eyes, wing veins, and hairs or bristles on various parts of the insects' bodies. There is no way to make this activity sound exciting and, unless I discover an undescribed species, it rarely is exciting. Although many of the species will be fairly common and easily identified, representative specimens must always be sent to a specialist on each group for confirmation of the identification. If I do not do this, I may find myself open to severe criticism later, particularly if I'm wrong: The middle of a homicide trial is not the best place to discover that I have misidentified a specimen; even if the misidentification is irrelevant to the estimation of the postmortem interval, doubt will have been cast on my capabilities and the reliability of my testimony.

At almost every site, there are a large number of species that belong to groups that are poorly known, are very difficult to identify, or are so small I can't actually see them without a microscope. With luck, I will be able to identify these specimens to the family level, but I will need to send them to a specialist for any identification to the genus or species level. The problem is to find such a specialist. Many groups of insects and other arthropods are not being studied by anyone in the United States, and some are not being studied anywhere in the world. Taxonomy, the description, identification, naming, and classification of organisms, has fallen out of fashion in the entomological community over the past 20 years, and now there are few people who are able to provide reliable identifications for many groups of insects. Even relatively common, well-known groups may present major problems. I submitted several specimens of a rove beetle in the family Staphylinidae to the Systematic Entomology Laboratory of the USDA in 1984. In 1996, I received a letter informing me that nobody there could complete the identification of the specimens because no one was currently working on that group. The laboratory did agree that the specimens were in the family Staphylinidae. I already knew the family, and most of the students in my general entomology class could have told me that in considerably less than 12 years.

Once the task of identification has been completed as far as possible, the specimens must be correlated as to species, stage of development, site of activity on the carcass, and time of occurrence. At this point the study once again becomes interesting as the relationships between the various species and the level of decomposition of the carcass begin to emerge. All this information is then correlated with the physical data collected from the body, internal and external ambient temperatures, daily rainfall at the site, relative humidity, and physical appearance of the carcass. I have to take all these factors into consideration when interpreting arthropods as evidence.

DECOMPOSITION IS A continuous process not amenable to division into discrete stages. Although recognizing this, most people conducting decomposition studies, myself included, have nonetheless tried to divide this process into stages because even though such stages do not exist in nature as distinct sets of physical parameters and assemblages of arthropods, they can serve as more or less uniform benchmarks for making comparisons between controlled decomposition studies and actual murder cases. This value becomes particularly apparent when one is faced with the task of explaining to a jury a series of events associated with a murder and the subsequent decomposition of the body.

Regardless of where decomposition studies are conducted, I have observed that certain patterns are common to most if not all of them. Although the species of arthropods involved differ from one geographic area to another, the same general groups are involved. These tend to arrive at the body in a predictable pattern, although the time frame may vary depending on the season and the environment. Over the years, a variety of different stages have been proposed, ranging from the nine stages of Megnin to only one stage posited in studies conducted by Cornaby on lizards and toads in Costa Rica and by R. E. Blackilth and R. M. Blackilth on mice in Ireland. My own work in Hawaii has shown that it is practical to divide decomposition into five stages. I established this division working with two of my graduate students, Marianne Early and Kate Tullis, during the mid-1980s. It seems to apply nicely to the data from most studies. The stages are Fresh, Bloated, Decay, Post-Decay, and Skeletal. I originally called the Post-Decay Stage the Dry Stage, but that was before we began working at a rain forest study site.

The Fresh Stage begins at the moment of death and ends when the body becomes visibly bloated. During this stage, decomposition causes few observable changes in the body's outward appearance. Aside from any wounds, the body closely resembles an immobile, unconscious but living person. The flies are not easily fooled, however. They quickly converge on the corpse, sometimes seeming almost to anticipate death. In Hawaii, I can expect the two most common species of blow flies to arrive within 10 minutes of death, day or night. (I realize that this statement flies in the face of conventional thinking on blow fly activity and may upset a few entomologists, since blow flies are generally thought not to be active at night, but I'll explain this in a later chapter.) The adult blow flies investigate the body and feed on any blood or secretions they can find. If the body appears suitable as a food source for maggots, the female blow flies will lay their eggs deep within the natural body openings and wounds. These eggs will hatch between 12 and 18 hours after they are laid, depending on the species of fly and the environmental conditions. The newly hatched maggots will immediately begin to feed on the tissues of the body.

Along with the blow flies, and occasionally even preceding them, come the flesh flies. Unlike the blow flies, the flesh flies do not lay eggs on the corpse. Instead the female keeps her eggs in her abdomen until they hatch and then deposits the maggots directly into the natural body openings or wounds. Frequently, the adult females do not even land on the corpse, instead squirting a stream of larvae directly into the opening or wound while still flying, not unlike a plane making a bombing run. This approach limits the number of larvae initially deposited, but unlike the blow fly eggs the flesh fly larvae can move around and avoid predators. And predators arrive at the body shortly after the adult flies. Wasps prey actively on the adult flies, often capturing them on the wing, and ants attracted to the corpse carry off both fly eggs and larvae. Later, various predatory beetles arrive at the corpse.

The Bloated Stage begins when the abdomen starts to show signs of inflation, but this point is hard to determine precisely. Immediately following death the processes of putrefaction begin. Any body, except for that of a newborn, has lots of bacteria in its digestive system. Most of these bacteria are anaerobic, not requiring oxygen for survival. In a living organism, the body defends itself against the activities of these bacteria, and as long as the bacteria stay within the digestive system, they cause no real harm. For example, in our intestines we harbor bacteria of the increasingly notorious species *Escherichia coli.* As long as these bacteria stay in our intestines, all is well. But if they invade other parts of the body, say the kidneys, severe complications and even death can ensue. After death, the body's defenses cease and the bacteria begin breaking the tissues down through the process of putrefaction. As a by-product of their metabolic activities, the bacteria produce gasses. And it is these gasses that cause first slight inflation of the abdomen and, later, the full inflation, that makes the corpse look somewhat like a balloon.

The corpse is still highly attractive to flies during this stage. Blow flies and flesh flies are still present in large numbers, and they are now joined by increasing numbers of house flies (family Muscidae). The greatest numbers of eggs and maggots are deposited during the early to middle portion of the Bloated Stage. The eggs are all laid at about the same time and so they hatch at about the same time, and the maggots congregate in large masses to feed. First they lacerate the tissues of the corpse with their mouth hooks and inject salivary enzymes that "predigest" the food before it enters their mouths. Combined with the activities of bacteria in the decomposing body, these enzymes reduce the food to a semi-liquid before it enters each maggot's mouth cavity. A mass of maggots is much more efficient than individual maggots at breaking down the tissue before ingesting it.

These maggot masses tend to remain intact and move through the carcass as a unit. The metabolic changes caused by the maggots and the anaerobic bacteria cause the internal temperature of the carcass to begin rising. During the Fresh Stage, the carcass temperature falls to near the ambient air and soil temperatures. Now the temperatures rise dramatically above these ambient temperatures, reaching highs as great as 53°C, or 127°F, much higher than normal human body temperature.

As the population of maggots increases and the maggots themselves increase in size, the corpse becomes even more attractive to various predatory species of beetles, ants, and wasps, as well as to parasites of the maggots and their pupae. The hister beetles (family Histeridae), rove beetles (family Staphylinidae), and burying beetles (family Silphidae) are the beetles most likely to be seen at this stage. Also arriving during this period are species of checkered beetles (family Cleridae). All these insects feed on the corpse or prey on the other arthropods present and lay their eggs on or under the corpse. A major site of activity during this and later stages of decomposition is the area of contact between the soil and the corpse.

During the Bloated Stage, fluids begin to seep from natural body openings and from any wounds present. These fluids, combined with the ammonia produced by the activities of the maggots, seep into the soil beneath the corpse, causing it to become alkaline. As soon as the body is placed at a site, the arthropods normally inhabiting the soil react very predictably—they leave. As the fluids from decomposition filter down into the soil, the area beneath the body becomes attractive to a series of organisms more specific to the decomposition process. Most of these organisms are microscopic and must be collected by processing soil samples in a Berlese funnel. The changes that occur in the populations of soil-dwelling organisms are long-term changes and become more and more significant as decomposition progresses. Regrettably, these are frequently the organisms that law enforcement agencies overlook when processing a crime scene.

The Decay Stage is the only other stage in decomposition besides the Fresh Stage that is marked by a discrete physical change. I consider the Decay Stage to begin when the combined activities of the maggot masses feeding externally and the anaerobic bacteria internally finally break the skin of the corpse, allowing the gasses to escape and the corpse to deflate. Large feeding masses of maggots are the predominant feature of the early to middle periods of this stage. The corpse is still moist then, and there are large amounts of decompositional fluids present on the surface of the soil and within the earth under the corpse. A distinct odor is associated with this stage of decomposition, and processing the corpse is not a pleasant experience. I have noted a strong correlation between the onset of the Decay Stage and a rise in absences and sick days among my graduate students.

In the Decay Stage there is a significant increase both in the number of individuals and in the number of species of beetles on the corpse. Both necrophagous and predatory species are present, and by the end of the Decay Stage, these beetles become the predominant insects on the body. The fly larvae usually have completed their development by the end of the Decay Stage. As the maggots near pupation, they stop feeding and enter a wandering, post-feeding stage when they leave the corpse and seek a place to pupate that is safe from predators. During the pupal stage, the maggot becomes immobile and does not feed. This stage of development is passed inside the hardened cuticle of the final larval stage. When it emerges, the insect has metamorphosed into the adult fly. As the maggots leave the corpse, the metabolic heat they generated during their development decreases, and the corpse's temperature begins to return to the temperature of its surroundings. During the Bloated Stage and the early periods of the Decay Stage maggots and beetles rapidly remove the flesh from the corpse. By the end of the Decay Stage, only 20 percent or less by weight of the corpse remains, consisting primarily of skin and bone.

As the corpse is reduced to skin, cartilage, and bones, the populations of blow flies, flesh flies, and house flies on the body decrease. They are replaced by other groups of arthropods, depending on the habitat. My early decomposition studies were conducted inside Diamond Head Crater, a dry area, and on the campus of the University of Hawaii at Manoa, which has moderate rainfall. In both these areas, the Post-Decay Stage was dominated by the adults and larvae of the hide beetles (family Dermestidae). Adults of these beetles begin to arrive during the later part of the Decay Stage, but larvae do not usually appear until the corpse has dried. The number of predatory rove beetles and hister beetles also increases during this stage, especially in the area surrounding the corpse. In the soil under the corpse, there is a steady increase in the numbers and variety of mites. Some of these—such as members of the families Acaridae, Histiostomatidae, and Winterschmidtiidae—feed on algae, fungi, and other by-products of decomposition, while others, such as members of the families Macrochelidae and Parasitidae, prey on insects and other small animals feeding on the corpse. Changes in the populations of these mites can yield valuable clues to the history of a corpse, particularly if it has been moved after death.

In my later decomposition studies, I began to investigate the decomposition patterns that occurred when the corpse was deposited in a wet area, such as a rain forest. It quickly became apparent that "dry" was not a term that could be applied to any phase of these studies. In fact, one of my graduate students became quite agitated at the thought of using the word "dry" to designate any part of her study, and we agreed to use the term "Post-Decay" for the fourth stage of decomposition. In wet habitats, such as rain forests and swamps, I have found that many of the beetles present in great numbers in other types of habitats simply do not occur. These include some of the necrophagous species, such as the hide beetles, and some predatory groups, such as the hister beetles. The absence of the hide beetles is

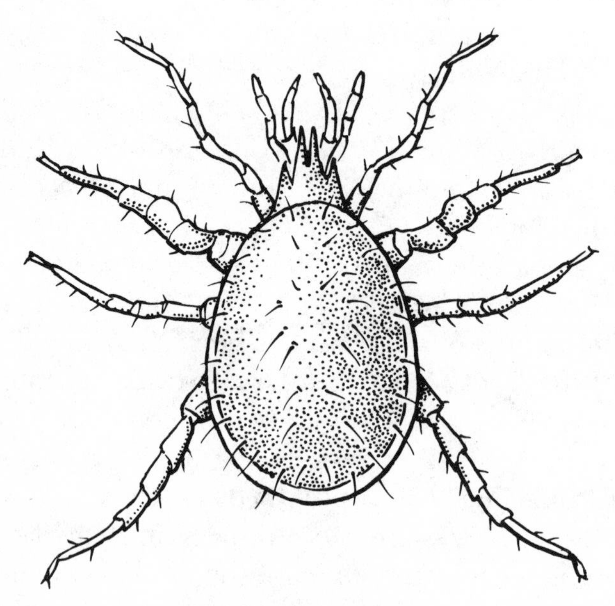

Adult Macrochelidae mite

understandable, because they require dried skin and cartilage as a food supply, and a corpse in a wet habitat simply does not dry out. The hister beetles do not require dried tissues as a food source, but still do not occur in wet habitats, at least in Hawaii. In wet habitats, beetles are replaced by other groups of insects, such as the nabid and reduviid bugs in the order Hemiptera. Maggots, however, stay on the corpse much longer than they do under drier conditions because the body tissues stay moist enough for them to eat. There are more species as well as additional generations of blow flies and flesh flies on a corpse in a wet habitat. Other groups of flies, such as the moth flies in the family Psychodidae, that are not normally found in drier habitats also exploit the corpse.

Regardless of the habitat, by the end of the Post-Decay Stage, only approximately 10 percent by weight of the corpse is left. Only bones and hair remain, and the corpse enters the Skeletal Stage. At this point, there are no insects on the body that are associated with decomposition. The remaining species associated with the body are very small, and live in the soil next to and beneath the skeleton. As time progresses, the normal soil-dwelling species gradually return to the area and those unique to decomposition depart. There is no definite end to this stage of decomposition, and there may be carrion-associated species present in the soil fauna for several months or even years after the death, depending on local conditions.

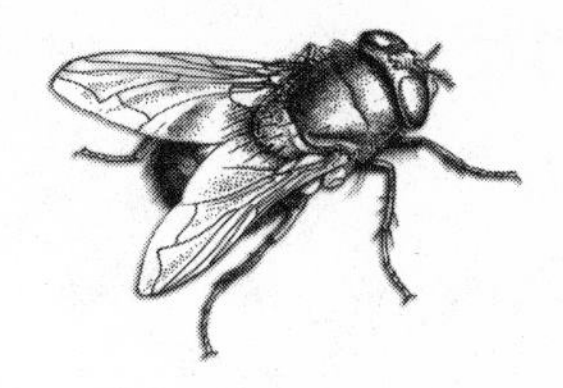

4

The First Flies

In my experience, for the first week or two of decomposition, the flies are usually the only uniformly reliable indicators of the minimum postmortem interval. Other insects visit the corpse during this period, but the times when they arrive are not predictable enough to be used in estimating the postmortem interval. The activity of these other insects may reinforce an estimate based on fly activity but does not, by itself, reliably indicate the interval since death. The flies, however, arrive early and begin their activities in a very predictable manner. The major challenge they present to the forensic entomologist is identification. As I noted earlier, it is difficult to accurately identify immature stages of insects, especially maggots. And although identifying adults in good condition is considerably easier, I and other forensic entomologists are all too frequently asked to work only with dead specimens, collected by crime

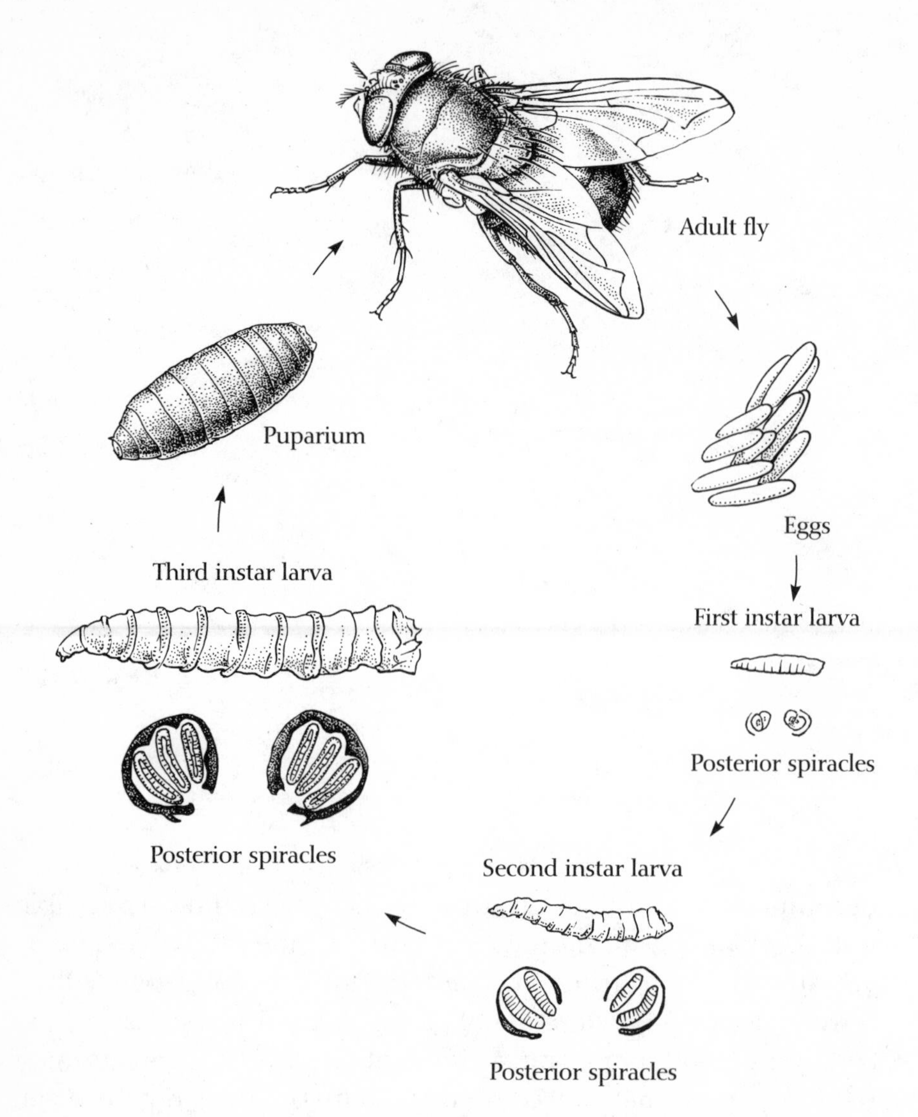

Life cycle of the blow fly *Chrysomya megacephala*

scene investigators and often submitted in marginal states of preservation. Naturally, such specimens tend to limit the accuracy of any estimate of the time since death.

To calculate the postmortem interval using flies, I must identify which stage of development each specimen represents. For most of the flies involved in decomposition, there are six distinct stages of development. The majority of these flies, including the blow flies and house flies and their relatives, lay eggs in masses in natural body openings or wounds. The time before hatching varies greatly, depending on the temperature and the species of fly involved. Research conducted in the state of Washington by A. S. Kamal in the 1950s gave developmental times for the egg stage of eight different species of blow flies. These studies were conducted at approximately 27°C, or 80°F, and the eggs required from 10 to 30 hours to hatch, depending on the species, with most requiring from 15 to 25 hours. The eggs hatch into a first stage maggot, called a first instar by entomologists. In the flesh flies, the other major group involved in the early stages of decomposition, the females deposit the first instar maggots directly on the corpse, but since each female is small she is not able to produce as many maggots at one time as the female blow fly eggs do.

Whether hatching from eggs or deposited as first instars, the maggots immediately begin to feed on the tissues and quickly increase in size. The maggot's external cuticle, composed of chitin, is flexible and provides protection from the environment, but it also limits the size of the maggot. Among insects and other arthropods, with few exceptions, new cuticle cannot be produced within a developmental stage. Once the maggot has grown large enough to fill the cuticle, it must molt—shed its cuticle—first producing a new, larger cuticle underneath the old.

Most of the flies involved in the early stages of decomposition have maggots of three stages, the first, second, and third instars. The first instar usually lasts the shortest time. Kamal studied a total of 13 species of blow flies and flesh flies, and for these the first instar ranged from 11 hours to 38 hours, with most of the species completing the first instar between 22 and 28 hours, and then molting into the second instar. The time required for development through the second instar for these same species ranged from 8 hours to 54 hours. For most of the species, this stage lasted about as long as the first instar stage. Twelve of the 13 species completed their development in 11 to 22 hours, and then molted into the third instar.

The third instar lasts the longest of the three stages and can be divided into two parts. During the first part of the third instar, the maggots actively feed on the corpse and maintain the tight maggot mass characteristic of maggots' feeding activities. This part of the stage lasts for 20 to 96 hours, according to Kamal. At the end of this time, the maggot has reached its maximum size and stops feeding. The second part of the third instar is the post-feeding, or wandering, stage. This is the longest period in the maggot's development. For the 13 species Kamal studied, this stage lasted from 40 to 504 hours, but most species took between 80 and 112 hours to complete this stage of development. During this period, the maggot's gut begins to empty as food is digested, and the maggot starts preparing for its pupal stage: it begins to decrease in length and, in most cases, eventually leaves the corpse to find a safe area for pupation, away from the predators and parasites attracted to the corpse.

Maggots usually move away from the corpse into a somewhat drier area, and for them, "drier" usually means "up." But moving up is not always a successful strategy. Once while working in Lyon Arboretum, a very wet habitat behind the University of Hawaii at Manoa, my students and I arrived early in the morning to find that maggots were leaving our dead pigs in search of a drier place to pupate. There was no dry area for several miles, but maggots have limited perceptions. They climbed trees. They crawled up the trunk, moved along the branches to their tips, and then fell back to the ground. Since there were three pigs in the area, each with thousands of maggots leaving to climb trees and eventually fall down, it was quite literally raining maggots. The deluge was so bad that we had to return to the laboratory for umbrellas so that we could finish our sampling.

Once the maggot has found a suitable place for pupation, it stops moving about and begins to form a pupa. At first, the color of the pupa is similar to the white to yellow color of the maggot, but over the next few hours, the cuticle darkens to a deep reddish brown. You can get some idea of the age of the pupa during this period from the color of the pupal case. At this stage the pupa looks like a small football, is resistant to heat, cold, drying, and flooding, and not very attractive to predators. While encased in the pupa, the insect metamorphoses into an adult, undergoing the equivalent of a second embryological development. The tissues and structures of the maggot are dissolved by a process called histolysis, and completely new structures—legs, eyes, wings, and so on—develop from groups of cells called imaginal buds or discs. This is a comparatively lengthy process and the time required is measured in days rather than hours. For the 13 species Kamal studied this stage lasted from 4 to 18 days, with most of the species completing development to adulthood in 6 to 14 days.

Most of the species of flies associated with the early stages of decomposition are in the group entomologists call the Muscamorpha, which includes the ubiquitous house flies, the blow flies, and the flesh flies. This group is also sometimes called the Cyclorrhapha, from the Greek for circular seam, because of the way the adult fly emerges from the pupa. There is a circular seam around one end of the pupal case, and the adult fly pops this "cap" off and then pulls itself out of the case. The newly emerged fly does not closely resemble the adult fly you are used to seeing. To be able to get out of the pupal case, the fly must be soft and pliable; its cuticle is light in color and soft, and its wings are wrinkled and collapsed. Over a period of several hours, the cuticle hardens, the fly assumes its adult coloration, the wings expand, and the fly becomes fully functional. The time required for development from an egg to an adult for the 13 species Kamal studied ranged from 10 to 27 days at a temperature of 80°F. These species had adult life spans ranging from 17 to 39 days and were able to begin reproducing 5 to 18 days after emerging from the pupae.

All these time spans for the stages of development and longevity are useful, but it is important to remember that they are derived from research conducted under controlled conditions. Outside the laboratory, the environment is variable, and forensic entomologists must take into account changes in such obvious conditions as temperature. Insects and other arthropods are not able to compensate for changes in temperature. Below a certain threshold, insects are too cold to fly and cannot lay eggs or deposit larvae. The rate at which the immature stages develop is also dependent on temperature.

As the temperature drops, the rate of development slows, until finally, at around 10°C, development ceases. The larvae do not die, they simply stop developing until the temperature rises to the point where development can resume. At even lower temperatures, most larvae die, but some insects can even withstand freezing. A few species of mosquitoes routinely lay their eggs in pools formed by melting snow; the larvae that hatch from these eggs are frequently frozen as ice forms in the pools, but they continue their development once the pools thaw.

Heat speeds development up. The higher the temperature, the more rapidly the larvae develop. But as the rate of development increases, the size of the adult flies decreases. I have seen this in studies conducted in my laboratory on two species of blow flies, *Chrysomya megacephala* and *Chrysomya rufifacies.* Smaller size was accompanied by a decrease in the number of eggs the females produced. So far, most of the species of flies I have reared in the laboratory had difficulty at constant temperatures around 35°C either in completing larval development or in reproducing once they reached the adult stage.

UNLESS THE CRIME scene is indoors, the corpse is exposed to constant temperature fluctuations during the course of an average day. The forensic entomologist must take this into account when estimating the postmortem interval from insect development patterns. In this respect, the Hawaiian Islands are a relatively uncomplicated area in which to work. The temperatures are fairly uniform from day to day and throughout the year.

There are variations, but nowhere as extreme as in many continental areas. Hawaii is also so close to the equator that the seasonal changes in day length are minimal.

But regardless of the geographic location, there are at least two major problems regarding temperature: First, the ambient temperatures at the crime scene virtually always differ from the temperatures used in laboratory studies to establish developmental times for different species of insects. Second, temperatures outside the laboratory fluctuate. So in calculating the postmortem interval, I need to account somehow for the differences in temperature between controlled studies and the fact that the temperature at the crime scene is constantly changing. The first step is to find out what the temperatures actually were at the crime scene during the period in question, usually the interval from the time the victim was last seen alive to the time of the body's discovery.

In the United States, temperature information is available from the National Oceanic and Atmospheric Agency (NOAA), which maintains a system of widely scattered weather stations. Most of these stations can provide hourly records of temperature, relative humidity, cloud cover, and rainfall. Other local weather stations may also provide weather information. In Hawaii, I have found that the pineapple and sugar cane growers are exceptionally cooperative in supplying me with data from their private weather stations. With large tracts under cultivation, these agribusinesses maintain an excellent network of weather stations over most of the islands. The military is an additional source of weather data, particularly for corpses that are found on or around air bases.

Even with such weather data, I face another problem. Dead bodies are rarely found close to weather stations and temperatures can vary greatly even half a mile from a weather station. In only one case I know of, reported by the late Ted Adkins, where the corpse was discovered under an airstrip weather station at the end of a small-plane runway in South Carolina, could the weather data be called truly accurate. In Hawaii, these variations are usually minimal, but they must still be considered. One way to solve this problem is to put a hygrothermograph where the body was discovered and record the daily temperature ranges for a week. I can then compare these temperatures with temperatures from the closest NOAA weather station and perform a statistical regression analysis. This analysis allows me to adjust the weather data to allow for differences between the NOAA station and the crime scene. Of course this analysis does not produce "real" temperatures the way a thermometer would, but it does give a more accurate estimate of the conditions at the site than weather station information alone. It's usually easier to perform the analysis than to obtain extensive weather data from the site. Once a crime scene has been released by the police, it often attracts not only the relatives and friends of the deceased and the news media, but also assorted curiosity seekers, some of whom may vandalize the site.

Once I have correlated the weather station data with the conditions at the crime scene, I am ready to begin estimating the postmortem interval based on fly development. In collecting insects from a corpse, I look for a representative sample of all the different species present on the body, trying to make sure that I collect the oldest immature stages as well as the adults of all the insects. During the early stages of decomposition, the most mature maggots or pupae usually will be the indicators of the minimum postmortem interval. These will have developed from the first eggs or larvae deposited on the corpse after death. To determine the species, I put representatives of each species into an environmental chamber in my laboratory and rear them to the adult stage. The other sample I fix and preserve.

For each species of fly I need to know the size of the maggot and which instar it represents. Using a dissection microscope, I measure the length of the maggot and then examine it carefully to determine its instar. A maggot has two sets of openings for respiration, the spiracles. One pair of spiracles is on the side of the third body segment near the head, and the other is at the tail end of the body. The posterior spiracles are the maggot's primary breathing structures and are frequently very distinctive in form. It is therefore often possible to make family- or even genus-level identifications based solely on the form of the posterior spiracles. Although the size of the maggot may give some indication of the instar, it is necessary to look at the posterior spiracles to be certain. In the first instar, the spiracles are a pair of relatively simple openings that may be shaped like a v or a u. In the second instar, there are two pairs of distinct openings. These are slits in the blow flies and flesh flies and somewhat serpentine formations in the house flies and their relatives. The openings are set in a pair of distinct, sclerotized structures. By the third instar, there are three well-defined openings on each side, and these are set in a pair of sclerotized plates or surrounded by a sclerotized structure called a peritreme. In addition to the posterior spiracles, the mouthparts may also help determine the instar. But since these are inside the maggot and some dissection is usually required to make them visible, the posterior spiracles are most frequently used.

Having determined the instar and length of the most mature of the maggots, I go back to the data from the laboratory rearings and see how long it took for a similar maggot to reach that stage of development under controlled conditions. I must then adjust the time period to fit the circumstances of the site where the body was found. There are several ways to do this. But the most widely used approach is to convert the temperatures and times into accumulated degree hours (ADH) or accumulated degree days (ADD) by multiplying the time by the temperature in degrees Celsius. Because the time required for development decreases as the temperature increases, the total number of ADH required to develop to any given stage remains constant. To get the time required to reach any given stage from the ADH for that stage, I simply divide by the temperature. For example, using data from the studies conducted by Kamal in 1958 at a temperature of 26.7°C (80°F), I can calculate the mode time required for the calliphorid fly *Phormia regina* to reach the third instar as follows. First, I add together the times required for completion of the egg stage, the first instar, and the second instar: 16 hours + 18 hours + 11 hours = 45 hours. To convert this to ADH, I simply multiply the time in hours by the temperature: 45 hours × 26.7°C = 1,201.5 ADH. And of course it's easy to convert the ADH to ADD: 1,201.5 ÷ 24 hours = 50.0625 ADD.

To see how the concept of ADD applies to actual cases, suppose that a body is discovered at 8:00 A.M. on October 15 and that insect specimens are collected and preserved at 9:00 A.M. the same day. The most mature *Phormia regina* maggots on the body are molting from the first instar into the second instar. The ADH required are the total ADH required for the maggots to complete the egg stage and all of the first instar. For laboratory rearings at 26.7°C, that time would be 34 hours, or 907.8 ADH. To estimate the period of insect activity, I work backward from the time when the maggots were collected, 9:00 A.M. on October 15. There were 9 hours of development between midnight and 9:00 A.M. on that day. The mean temperature at the scene for that period was recorded as 20°C. This means that a total of 180 ADH were accumulated on October 15 (9 hours × 20°C = 180 ADH). On the previous day, the mean temperature was 21°C, and this gives a total of 504 ADH (24 hours × 21°C = 504 ADH). Adding the totals for the two days, I get 684 ADH. Subtracting this from the 907.8 ADH required leaves 223.8 ADH. The mean temperature on October 13 was 20°C, with each hour of that day accounting for 20 ADH. When I divide 223.8 ADH by 20°C, I a total of 11.2 hours of development that would have occurred on October 13. Counting backward from midnight, I arrive at an estimated starting point for insect activity between noon and 1:00 P.M. on October 13. This may not be the actual time of death, but it is the minimum period of time that could have elapsed between death and collection of the insects.

Once I have determined from laboratory rearing data the total number of ADH required to reach the most mature stage of development I collected from the body, I simply work backward from the time I stopped the biological clock by collecting and preserving the insect specimens until I reach the total number of

ADH. This time marks the onset of insect activity. In making these calculations, I can use a mean of the temperatures recorded from the site on an hourly or a daily basis, or I can calculate the ADH values for each of the hourly temperatures given in the weather reports. Overall, mean temperatures usually provide a more accurate assessment of the actual conditions under which the maggots developed than hourly temperatures. With the hourly approach, I must assume that the temperature of the corpse changes as quickly as the temperature of the surrounding air, which is not true except in very unusual circumstances. Normally, the temperature of a solid changes more slowly than the temperature of the air around it (a piece of steak put in a freezer, for example, does not immediately freeze; some time passes while its temperature decreases until finally the steak freezes); and long-term mean values therefore usually prove to be more accurate.

Another significant factor I must consider when using ADH or ADD calculations to estimate the postmortem interval is the heat generated by the maggot feeding mass. In many cases, ambient air temperatures do not accurately reflect the temperatures at which the maggots develop inside the corpse. In my decomposition studies in Hawaii, I have found that internal carcass temperatures are elevated by maggot mass activity up to 22°C above ambient air temperatures. At first glance, this fact seems to be good grounds for not using ADH or ADD calculations in estimating the time required for maggot development. But the ADH and ADD calculations do work. Although heat generated by the maggot mass does influence the rate of development, this heat is not generated immediately. It may take several days to develop. This lag may be the result of the lack of an organized compact mass at the early instar stage. The amount of heat generated is also dependent on the size of the maggot population. In the case of bodies found during relatively cool weather with small populations of maggots, ADH or ADD calculations will be more accurate than they are for bodies found in warmer weather with more maggots present.

Moreover, the temperature of a maggot mass does not necessarily reflect the actual temperatures at which the maggots are developing. The temperature of a mass is recorded at the middle of the mass, and temperatures are lower on the outer portions of the mass and in other parts of the corpse. Since maggots cannot regulate their own temperatures, at ambient temperatures above 50°C (122°F), they are in danger of thermal death, and thus they do not spend much time at the center of the mass. Instead they seem to circulate throughout the mass, moving to the inside to feed and, as their temperatures become dangerously high, moving back to the outside of the mass to cool down and digest. After a period of cooling, they reenter the mass and repeat the cycle. In the process, they spend a good deal of their time at temperatures lower than the temperature at the center of the maggot mass.

A factor frequently overlooked in estimating the postmortem interval is the design of the studies from which forensic entomologists obtain baseline data on maggot development. In the vast majority of these studies the researchers tried to create an ideal environment for maggot growth. This means raising large numbers of maggots in relatively small containers—in other words, maggot masses. Although there are a few studies of individual maggots, they are almost never used in forensic entomology. Work conducted in my laboratory by one of my graduate students showed that under laboratory conditions the temperature rises even with small population densities of maggots. So the studies on which forensic entomologists base their calculations are already, if unintentionally, providing some adjustment for temperatures generated by the maggot mass.

In addition to influencing the rates of development, temperatures may also limit the species of maggots that feed on a corpse. Some species of blow flies cannot tolerate the high temperatures within a maggot mass and may be excluded from that portion of the corpse infested by masses or from the corpse entirely. Careful sampling must include collecting specimens from all areas where insects are present. Usually there are quite a number of species present after the first several days of decomposition. And so whenever possible, I use the ADH or ADD calculations for as many species of maggots as possible in estimating the postmortem interval. Agreement between the individual developmental times makes the estimate more accurate.

Although I have emphasized the significance of maggot development so far, the pupal stage is also important. A fly spends as much as 40 percent of its life cycle as a pupa, and a good deal of evidence can be obtained from pupal cases. Forensic entomologists have recently begun to devise various methods for determining the age of pupae, including dissections to determine the age of the developing fly inside and assessment of the color changes in the pupal case that occur during its early development. In most of my cases, I simply placed the pupa into an environmental chamber at a known temperature and noted how long it took for the adult fly to emerge. Using ADH or ADD calculations with temperatures from the crime scene and the known temperatures from the environmental chamber, I can estimate the time between when the eggs or larvae were deposited and when the adults emerged. Of course in providing the final estimate I must take into account the time spent in the environmental chamber. I must also examine the corpse's nearby surroundings for empty pupal cases; if any are present, at least one generation of flies has completed development and left the corpse. In such a case, any maggots present on the corpse will be from a later egg or larval deposition.

HERE'S HOW THIS type of determination of the postmortem interval worked in a case referred to me by the U.S. Air Force Office of Special Investigations (AFOSI) in the spring of 1992. On May 5 the investigating officer asked me to assist in estimating the postmortem interval in the case of a 15-year-old military dependent found dead on March Air Force Base, near Riverside, California, at 7:00 P.M. on April 28, 1992. At the scene, investigators noted that the victim, a male, had multiple stab wounds to the chest and that his throat had been cut. The body was removed by personnel from the Riverside County Coroner's Office and put into a freezer at 10:30 P.M. The temperature in the freezer was held at 18°F, low enough to prevent any further development by the maggots on the body.

When the autopsy was conducted, on Friday, May 1, the regional forensic consultant collected samples of insects. He also collected soil samples at the scene from the area under and immediately adjacent to where the body had lain. This consultant had attended workshops on forensic entomology and he followed the procedures outlined in those sessions. He collected and packed separately the insect specimens from each part of the body, and split each of the three batches of insects into two sublots. One was fixed with hot water and then preserved in a solution of isopropyl alcohol and water to prevent hardening of the maggots. The other sublot was put into a cardboard box with some beef liver for food and some fine sand to provide protection for the maggots. The soil sample was also put into a cardboard box.

The specimens were not shipped to me until Thursday, May 7, and since the weekend was approaching, I asked to have the specimens sent via Federal Express to my home address. I did not want to wait until the beginning of the week to start working on them. As soon as the shipment arrived, I took it to my laboratory at the university. There were 46 crime scene photographs, hourly weather data from the base for the period from April 21 to April 28, 1992, insect specimens in six separate containers and the soil sample. I put the live specimens into the environmental chamber at 26°C. After measuring the preserved maggots and determining which instar they were in, I transferred the preserved specimens to 70 percent ethyl alcohol for storage.

The photographs of the crime scene showed the victim lying face down in a field of grasses and weeds. The grasses were typical of the vegetation that blow flies seek out at night. Four of the photographs showed enough detail for me to identify a specimen of a rove beetle, *Creophilus maxillosus.* This species of beetle preys on the eggs and maggots of blow flies and flesh flies. The victim's face was obscured by a mass of maggots, as was the area around the neck wound. There was a lot of blood on the chest, but it was not heavily infested with maggots, probably because the prone position of the body had prevented most flies from reaching it. There were egg masses along the sides of the body, where it had come into contact with the soil.

I identified both blow fly and flesh fly maggots in the samples submitted. The flesh fly maggots entered the pupal stage but did not develop into adults, not unusual for maggots in this family when they are refrigerated. Considering the location and the size of the maggots, the specimens probably were *Sarcophaga bullata,* a carrion-invading species common in the southwestern United States. I identified three species of blow flies in the sample, second and third instar maggots of *Phaenicia sericata,* third instar maggots of *Phormia regina,* and third instar maggots of *Chrysomya*

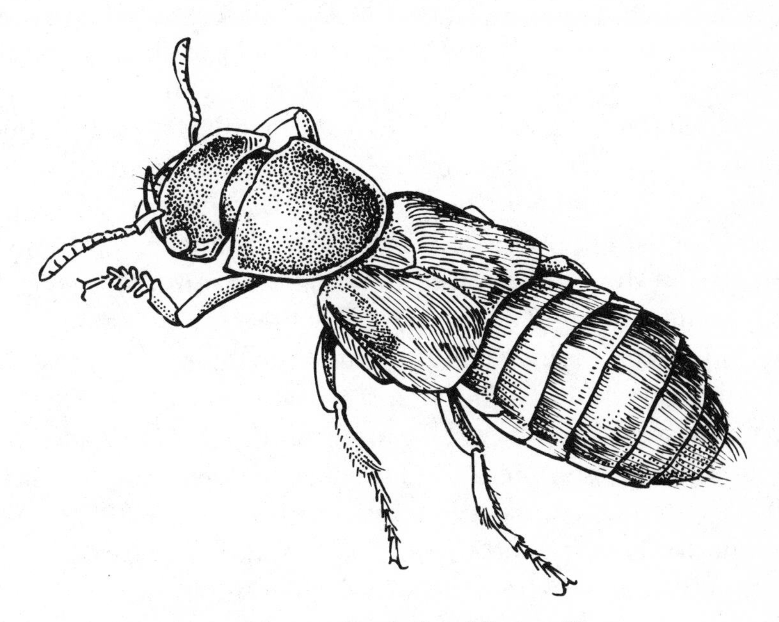

Adult of the rove beetle *Creophilus maxillosus*

rufifacies. The largest third instar maggots of *Phaenicia sericata* were 11 millimeters long and, according to laboratory rearing data, would have required a total of 1,689 ADH to reach that stage of development. I used both hourly and mean daily temperatures to calculate the ADH values that would have been required by *Phaenicia sericata* maggots at the site. Both calculations indicated that between 72 and 77 hours had elapsed between the death and the refrigeration of the body at the morgue. Using the hourly temperature data provided by the base, I concluded that the most probable time of death was between 10:00 P.M. and midnight on April 25, 1992. The specimens of *Chrysomya rufifacies* indicated the same time frame. But those of *Phormia regina* had undergone a shorter period of development. This was not surprising because *Phormia regina* frequently does not invade a corpse immediately after death. Thus the shorter developmental period for this species was not inconsistent with the known facts.

My only problem at this point was that my estimate required the blow flies to lay eggs at night. For years it has been widely accepted that blow flies cease all activity at night. During some of my early decomposition studies in Hawaii, I wanted to set my test animals out as early in the morning as possible. To do so, I sometimes went to the study site while it was still dark, and on these occasions I was surprised to see blow flies arriving almost as soon as I put out the carcasses. Later I observed blow flies around crime scenes during periods of darkness. Then in 1990 Bernard Greenberg published the results of some experiments he conducted in the Chicago area showing that some blow flies lay eggs quite late at night. Although blow flies do not usually fly or seek food or egg-laying sites at night, if the temperatures remain above the threshold required for activity and a suitable food source is dumped nearby, they can become active and lay eggs. (I have observed, however, that blow flies lay fewer eggs at night.) In the case of this corpse, the temperatures were well above the threshold for activity and the area where the body was left was typical of the places where blow flies rest at night.

As soon as I completed my analyses on May 14, I telephoned

the Office of Special Investigations and gave the investigator my results. In the written report, I gave my estimate in terms of days since death, but I also told him on the phone that I thought death had occurred between 10:00 P.M. and midnight on April 25, 1992. There was total silence at the other end of the telephone line. After I said "hello" a couple of times, the agent asked me to hold the line and repeat my findings to another investigator. I anticipated an outburst of laughter or some other indication that I was completely off base for some reason or another. Instead, after I repeated the estimate to the second agent, there was another long silence. When speech returned to the other end of the line, I was informed that they had picked up a suspect and were interrogating him when I called. While I was calling in my estimate of the time of death, the suspect was confessing to the murder. He said that death occurred at approximately 10:30 P.M. on April 25, 1992.

I WISH I COULD say that all cases are resolved this neatly, but most are not. As a rule, the more time elapses between death and the discovery of the body, the less accurate the estimate of the postmortem interval based on entomological evidence. And even in some cases where the time between the victim's last being seen alive and the discovery of the body is relatively short, other factors may diminish the accuracy of the estimate. In a case in Wyoming, weather proved to be the complicating factor. The victim had been dishonorably discharged from the air force on drug-related charges on April 15, 1994. He was last seen alive at approximately 11:00 A.M. in Layton, Utah, just outside the gates to the air force base. The last person to see him alive later became the primary suspect in the investigation. The body was discovered by hikers at approximately noon on April 17, 1994, just outside the small town of Carter, Wyoming (population: 5). The Uinta County sheriffs deputies called in the Wyoming State Crime Lab to process the scene. That team could not get to the site until the next morning, so the body was left in situ overnight.

During the autopsy, conducted in Fort Collins, Colorado, the pathologist observed that the victim's hands were handcuffed behind his back. There were abrasions to the wrists that were consistent with these bonds, as well as a few bruises to the hands and elbows. And there was an incised wound on the victim's chest that the pathologist concluded might be consistent with the victim's clothing being cut from his body. There was also a fracture to the hyoid bone, which might be associated either with manual strangulation or with a blow to the neck. The investigating agent requested that the potassium levels of the vitreous humor from the victim's eyes be analyzed to determine the approximate time of death, but he was told that the body was too decomposed for such an analysis to be effective. Insect eggs and larvae were present around the victim's mouth and genitals, but none were associated with the incised wound on the chest.

On April 26, 1994, a representative of the Air Force Office of Special Investigations asked me to provide an analysis of the insect evidence. I agreed and the evidence was shipped to me via U.S. Express Mail.

On May 3, I received the materials: two vials containing the insect specimens, weather data from three weather stations close to the crime scene, 25 video stills from the crime scene, a transcript of an interview with the pathologist who conducted the autopsy, maps showing the location of the crime scene, and a video of the scene. Although a video record of the crime scene can be invaluable when you are analyzing a scene you are unable to visit yourself, it has its limitations. The swaying of the camera can be disorienting, and a video does not provide the level of detail that still photographs do.

One of the two vials contained specimens preserved in isopropyl alcohol: eggs and a few first instar maggots. These maggots appeared to have just emerged from the eggs and, in fact, some of them were still partially inside the eggshells. It seemed probable that these maggots had hatched in response to being put into the isopropyl alcohol. The second vial contained maggots on gauze and some eggs. The eggs were mature and ready to hatch, but had become desiccated following their collection. The maggots had begun to develop, but had also dried out following collection. Along with this vial there was a record of observations from the laboratory of the maggots' growth from April 20 to April 26, the day they were shipped. The entry for each day was the same: "No Change." Since no food had been provided and maggots do not thrive on a diet of gauze, I presumed they had been dead almost the entire time. Identification of these maggots and eggs was severely impeded by the condition of the specimens. They were a species in the blow fly genus *Phaenicia,* but I could not make any further determination from only dehydrated first instar maggots and eggs. Two species are commonly found in the area: *Phaenicia sericata* and *Phaenicia coeruleiviridis.*

Although the species identifications were in doubt, there was still the possibility of making an estimate based on the time needed for the eggs to develop and hatch into first instar larvae. To provide an estimate of this period, I had to analyze temperature data supplied by weather stations in the vicinity of the crime scene. Temperatures for the period in question, April 15 to 18, 1994, ranged from 6° to 24°C, with an overall mean temperature of 16°C. Development from egg to emerging first instar larvae at these temperatures would require just over 24 hours for species in the genus *Phaenicia.* Thus the period required for development to the first instar of eggs laid on the body would be over 24 hours. Further, species of *Phaenicia* prefer to lay eggs on sunny days when the ambient temperatures are above 21°C. The maximum temperature recorded for April 15 was 18°C, below the usual egg-laying threshold. On the two following days, temperatures were at or higher than the threshold of 21°C, with highs of 21°C on the sixteenth and 24°C on the seventeenth. Temperatures suitable for egg laying occurred between noon and 7:00 P.M. on both days.

If the body was at the scene on April 16 and eggs were laid, the eggs would all have hatched into first instar maggots, which would have been actively feeding and possibly have developed into early second instar maggots by the time the specimens were collected on the morning of April 18. If the body had been available for egg laying during the warm period of the day on April 17, the eggs would have had just enough time to hatch into first instar maggots on the morning of the eighteenth. I thought that this scenario was consistent with the condition of the eggs and larvae given to me for examination. In my report, I said that the eggs were most probably laid during the afternoon of April 17, 1994.

In this case, the activity of insects on the body did not account for the period of time between the last time the victim was seen alive and the discovery of the body. The corpse was also more decomposed than one would expect if the body had been exposed for only the period of approximately 24 hours indicated by the insect activity. Clearly, some aspect of the case was yet to be explained. One thing I found bothersome was the absence of any fly eggs or maggots associated with the wound on the victim's chest. Although natural body openings provide the most frequent access for maggots, a chest wound with its fresh blood should also have attracted species of *Phaenicia.* One possible explanation was that the body had for some reason not been accessible to the adult flies until some time had elapsed. If so, the blood would have dried and the wound would not have been as attractive to the flies for egg laying.

When the case was finally solved, this proved to be a correct assumption. The murderer had killed the victim on April 15, and then stashed the body in the trunk of a car, closed the trunk lid, and kept it there until he dumped it at the site on the morning of the seventeenth. Flies had laid their eggs on the body during the warm periods of the day on the seventeenth, and the mature eggs and hatching maggots had been collected on the morning of the eighteenth. The insects could not get to the body to lay eggs while it was inside the trunk of the car, and in the elapsed time the blood from the chest wound dried, becoming less attractive for egg laying when the body was exposed.

My estimate of the period of insect activity in this case was quite accurate, but it was not an accurate estimate of the postmortem interval. In some cases, there are factors affecting the onset of insect activity that are not readily apparent from the information available when the entomologist is analyzing the evidence. This fact is frequently overlooked by both the legal system and, unfortunately, entomologists just beginning to work in forensics. I always try to emphasize that what I produce is an estimate not of the postmortem interval, but of the time insects have been active on the corpse. Frequently the two are very close, but sometimes the period of insect activity is significantly different from the postmortem interval. The entomological estimate must be considered along with the other facts in the case. The insect evidence is only a part of the total picture, frequently a very significant part, but not by itself the whole picture.

5

Patterns of Succession

After the first two weeks of decomposition, the blow flies and the flesh flies start to leave the corpse to pupate, and since they do not usually return to the same corpse to produce a second generation, their usefulness as indicators of the minimum period since death decreases. After the departure of these flies, the emphasis in estimating the postmortem interval shifts from the developmental cycles of individuals and species to the succession patterns of all the insects and other arthropods present on and around the corpse during the various stages of decomposition.

As the maggots of the blow flies and flesh flies remove the moist, soft tissues of the body, the corpse begins to dry and to attract such species as the hide beetles, which eat dried skin and cartilage but do not like moist food. Some species that prey on the maggots are not equipped to prey on the better-armored beetles or their larvae. And so the predators and parasites that exploit flies leave the corpse as the maggots disappear, being replaced by species that are able to feed on the beetles and their larvae. Some species of beetles arrive as adults and lay their eggs; then their developing larvae prey on the maggots. Their rate of development is usually keyed to the rate of development of the maggots and they reach the pupal stage when the maggots depart. The larvae and adults of many insects are equipped to use completely different food sources. Thus the predatory larvae of a species that feeds on maggots may develop into adults that feed only on the dried tissues of the corpse. Unless interrupted, patterns like this continue, although with a succession of different players, until the corpse is reduced to skeletal material and the normal fauna of the area returns. Months or even years may pass before the area around the corpse returns to normal.

To use succession patterns successfully when I am estimating the postmortem interval, I must rely heavily on the data obtained from decomposition studies that I have conducted in various habitats—or for cases outside Hawaii, on decomposition studies done by others. Ideally, of course, I would have such data for the exact spot where the body was discovered. This has happened to me only once; in that case, the corpse was dumped into a ravine just outside of Lyon Arboretum in the Manoa Valley, approximately 25 meters away from where I had been conducting decomposition studies. Almost always, however, the corpse is discovered some distance from the location of any of my decomposition studies and I must find the best match I can with known sites.

Even though estimations of the postmortem interval during the early stages of decomposition are based primarily on the developmental rates of individual species of flies, succession patterns do enter into these estimates because most species of flies lay eggs more than once on a dead body. Typically the corpse is attractive to female flies of some species for several days. After that period, the corpse will have changed and will no longer be an attractive egg-laying place for these species of flies, and other species will take over. But since there have already been several days of egg laying before this switch occurs, there will be several different stages of development of any given species of maggot on the corpse after the first day of insect activity. As decomposition progresses, these different stages will continue in waves, ending only when the maggots from the last clutch of eggs laid complete their development to the pupal stage. The presence or absence of these different stages on a corpse can buttress the estimate derived from the development of a single set of maggots.

CONSIDER THE CASE of a body discovered by a jogger at the edge of Kawainui Marsh on Oahu at approximately 5:30 P.M. on August 26, 1985. When the police arrived, they found the body of a male, clad in a T-shirt and pants, lying supine at the edge of the marsh with legs extended and arms at the side. The body was decomposed beyond recognition and heavily infested with maggots. The lower abdomen had been opened by maggot activity and the pants were pulled down to mid-thigh, but still fastened. A hat, with what appeared to be a bullet hole above the visor on the right front, was found near the body; it was later shown that this hat belonged to the victim.

I first saw the corpse during the autopsy at the morgue on the morning of August 27. Upon entering the autopsy room, I immediately noticed a strong odor of ammonia. There was also a greenish discoloration of the skin, which I have since learned to associate with a body that has been immersed in water. The head had been stripped to the skull, although some skin clung to the sides and the ears were largely intact. The upper portions of the chest were skeletonized and contained a large mass of late instar maggots. The groin area was also largely decomposed and contained both early and late instar maggots. Maggots had not yet invaded the abdominal cavity, which was still intact. There were also maggots on the arms and legs, but they were not forming feeding masses. The cause of death was said to be a gunshot wound to the head, and the medical examiner declared the death a homicide, but no fragments or intact bullets were recovered from the body.

After collecting and treating my specimens I returned to my laboratory with the maggots. Initial examination of the maggots showed that there were at least two species present, each represented by several different stages of development, indicating that adult flies had laid eggs at several different times. I easily identified one of the species as *Chrysomya rufifacies* because this blow fly's maggots have quite distinctive spinelike projections and stand out from the rest of the maggots found in Hawaii. This species was present in a large mass in the chest cavity and in the groin area, and on the arms and legs. The specimens I collected from the chest cavity and groin were second and third instar maggots; those from the arms and legs were post-feeding third instar larvae, measuring 12 to 15 millimeters in length. These maggots, found as both second and third instars, lacked the distinctive spines of *Chrysomya rufifacies.* The most mature third instar maggots for this species were 14 to 16 millimeters long. Although this species was abundant on the chest and in the groin, I did not find any specimens on either the arms or legs. I placed samples of both the second and third instar maggots on beef liver and reared them to the adult stage in the environmental chamber. I suspected that these would prove to be the other common species of blow flies found during the early stages of decomposition in Hawaii, *Chrysomya megacephala.* When the adult flies emerged, my suspicion was confirmed.

Using data from life history studies of both species I had conducted under controlled laboratory conditions early in my career, I determined that it would require approximately 2,820 ADH for *Chrysomya rufifacies* to reach the stage of a 15-millimeter-long post-feeding maggot and 2,725 to 2,939 ADH for *Chrysomya megacephala* to develop to a maggot 14 to 16 millimeters long.

Since the two species arrive at a corpse at approximately the same time, this made sense. Usually, *Chrysomya megacephala* arrives a little before *Chrysomya rufifacies,* and I might have anticipated that a few more mature *Chrysomya megacephala* maggots would have been present. One explanation for this lack of larger specimens of *Chrysomya megacephala* maggots might be that *Chrysomya rufifacies* maggots tend to become predators during the later stages of their development and their favorite prey seems to be *Chrysomya megacephala.* To adjust the laboratory data to fit the conditions where the body was discovered, I used temperature data from the weather station at the Kaneohe Marine Corps Barracks less than 2 miles from the site. The temperatures from this station ranged from 24° to 26°C during the period in question. These values resulted in a postmortem interval estimate of approximately 5 days before the discovery of the body.

In addition to the time required for development of the most mature maggots collected, I also took into consideration the different stages of development represented by the maggots of both species. I compared these stages with the results of decomposition studies conducted in a habitat on Oahu similar to that where the corpse was found. In those studies, on the fourth day, there were first, second, and third instar larvae of both *Chrysomya megacephala* and *Chrysomya rufifacies* on the body. On the fifth day, only second and third instars of both species were present, and during the second sampling period of the day post-feeding third instar maggots of *Chrysomya rufifacies* were collected. By the sixth day, only third instar maggots of both species were present on the body. All these data supported the estimated postmortem interval of approximately 5 days before discovery of the corpse.

When the victim was identified, it was clear that the estimated postmortem interval fit the circumstances well. The man had last been seen alive by a relative 5 days before the discovery of the corpse. He had left home for work at approximately 6:00 P.M. and failed to report for work at 8:00 P.M. as scheduled. A suspect was identified and charged with second-degree murder, and he was subsequently convicted when the trial was held in 1989.

ANOTHER CASE DEMONSTRATING the importance of the numbers and kinds of insects that succeed the flesh flies and blow flies is that of a corpse discovered in a pineapple field just off the H-2 freeway in Waipio, on Oahu. I was away from the islands on a trip to the mainland when the body was found and could not visit the scene before the body was removed. I did examine the corpse in the City and County Morgue when I returned on the morning of October 16, 1989, and made my collections. The body was in a fairly advanced state of decomposition, and at first even determination of the victim's gender was difficult. The corpse appeared to be between the Post-Decay and Skeletal Stages of decomposition. At the time I made my examination, the corpse was clothed in a tank top and a pair of shorts.

I was struck by the variety of insects on the corpse. Above the right eye, I found a mass of empty pupal cases of the blow fly *Chrysomya rufifacies.* By the seventeenth day of decomposition in an open pineapple field, all of the specimens would have completed their development to the adult stage and flown away, leaving behind only empty pupal cases; so I was sure that more than 17 days had elapsed since death. I saw several other families of flies on the corpse. I collected third instar larvae of a species of flesh fly that were 15 to 16 millimeters long. Occasionally, flesh flies will deposit larvae on a corpse in the later stages of decay, especially if the surroundings are wet, as they were in this case; so the presence of these larvae was not totally unexpected.

I also found larvae of the cheese skippers on the corpse. After rearing the larvae to the adult stage, I could identify them as *Piophila casei,* a species commonly found on decomposing remains in Hawaii until approximately 36 days after death. The specimens I collected from the corpse entered the pupal stage within 1 day of being placed in the rearing chamber. So I moved the estimated time since death forward to slightly over 1 month. Also on the corpse were maggots of a species of fly in the family Otitidae, the picture-winged flies. These flies are also common during the later stages of decomposition but their maggots are not usually found on a corpse past day 37. One other species was of particular significance: the black soldier fly, *Hermetia illucens,* which was represented by maggots 10 to 14 millimeters long. This fly does not usually become attracted to a decomposing body until at least 20 days after death; so the presence of such large maggots also indicated a postmortem interval of approximately a month. Given my previous experience with this species in cases and decomposition studies, I suspected that these were not the most mature specimens associated with the corpse. Those would probably have been in the soil under the corpse and might still be at the scene.

The beetles on the corpse also indicated an extended postmortem interval. There were both adults and larvae of the hide beetle *Dermestes maculatus* feeding on the dried skin and cartilage.

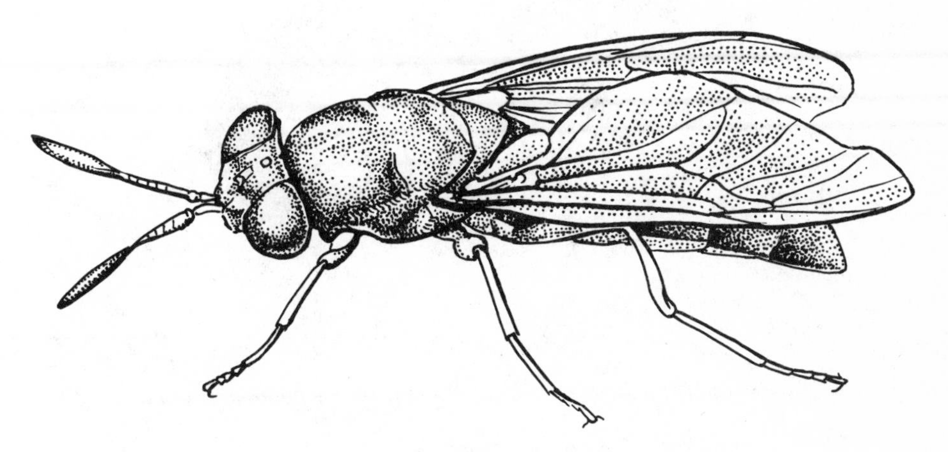

Larva (below) and adult (above) of the black soldier fly, *Hermetia illucens*

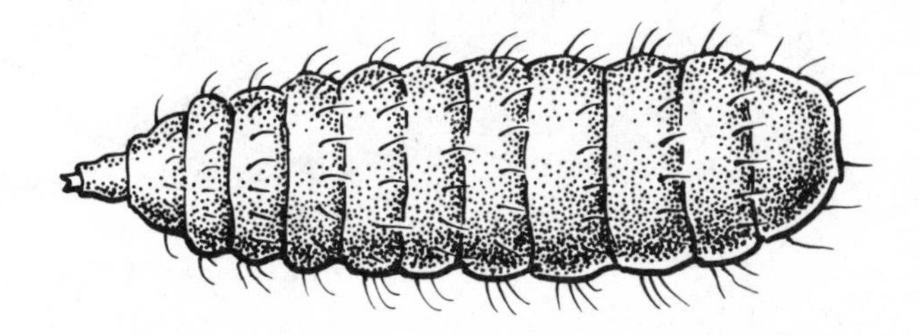

Adults of two species of checkered beetles were also presents. The most numerous of these were adults of the red-shouldered ham beetle, *Necrobia ruficollis,* a species I frequently find on corpses, but not usually in large numbers. The other species, the red-legged ham beetle, *Necrobia rufipes,* was also there in large numbers, as is typical for an exposed corpse about a month after death. Since I could see no larvae of either species on the corpse itself, I thought these might still be in the soil. There were also adults of a single species of rove beetle, *Philonthus longicornis,* on the corpse, but no larvae.

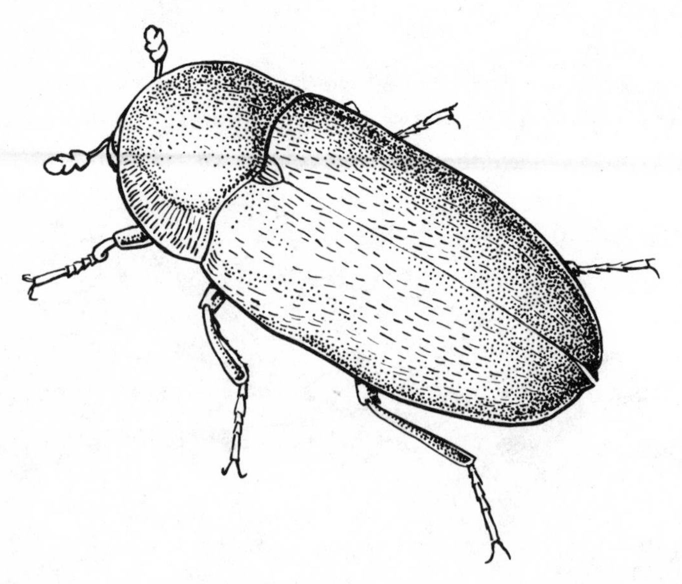

Larva (below) and adult (above) of the hide beetle *Dermestes maculatus*

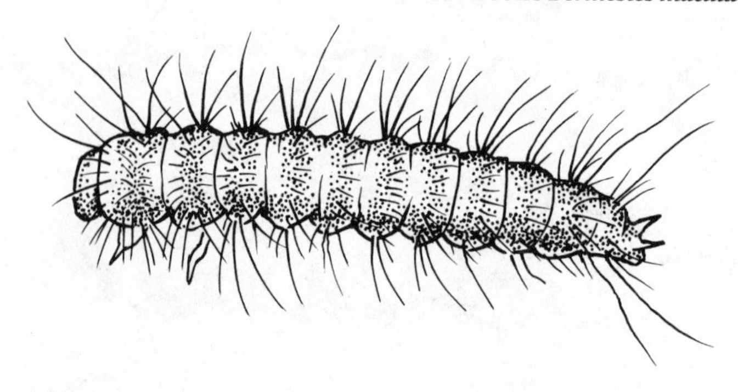

A surprising addition to the corpse fauna were a number of immature and adult crickets. These were identified for me as *Teleogryllus oceanicus* by a graduate student in the Department of Entomology who was working toward his doctoral degree on systematics of Orthoptera. I had only rarely encountered crickets on corpses and never before in numbers. Most crickets are omnivorous, and thus it was not surprising to find them their feeding on a corpse, but the numbers in this case were remarkable.

Overall I concluded from my examination of the corpse at the morgue that I was looking at a postmortem interval of slightly over a month, and also that a significant part of the insect fauna had probably not been collected and still remained at the scene. Naturally, there would have been some migration away from the scene once the corpse was removed. But with approximately a month of decomposition, there should have been enough decompositional fluids seeping down into the soil to provide a food source for many of the insects left behind. I definitely needed to visit the scene as soon as possible. With the assistance of the lead detective on the case, I was able to arrange for a supervisor from the pineapple company to meet me at the scene on the afternoon of the sixteenth. This supervisor had been present when the corpse was recovered and could guide me to the exact spot with no difficulty. The site was near the edge of one of the service roads leading into the field and was easily accessible from the main road. The pineapple plants at the site were trampled, probably by the investigators, and discolored, probably by the decompositional fluids leaking from the corpse.

Once at the scene, I could easily tell where the corpse had lain. The soil had a color and texture different from those of the immediate surroundings as a result of the seepage of fluids from the corpse during decomposition, and there was still a strong smell of decay. I immediately noticed large numbers of crickets in the area. These proved to be both adults and immatures of the same species I had collected from the corpse at the morgue, *Teleogryllus oceanicus*. I also collected from the soil a number of rove beetles representing several different species.

Several empty pupal cases of the blow fly *Chrysomya rufifacies* were scattered on the surface of the soil immediately adjacent to the spot where the corpse had lain. These were consistent with the typical pattern for this species—some individuals undergoing pupation on the corpse and the rest migrating away from the corpse to pupate on the soil surface. I also collected adults of the scarab beetle *Aphodius lividus* from the soil in the area where the corpse had lain. As I anticipated, there were a large number of maggots of the black soldier fly, *Hermetia illucens,* in the soil, and these appeared to be in a later stage of development than those I had collected at the morgue. Many of the specimens I thought might be present at the scene are usually found under the surface of the soil rather than on top. To collect these, I took soil samples from where the corpse had lain and the immediately adjacent area, watched by the bewildered pineapple plant supervisor.

On my return to the laboratory in the late afternoon, I put the soil samples into Berlese funnels to extract the arthropods. At the time, I had a small room assigned to my laboratory that had been specifically designed for Berlese processing. It was supposed to be vented directly to the outside of the building rather than connected to the central air conditioning system for the building. At least that was what the building plans said. But by 8:00 A.M. the next morning, it was apparent from a pervasive odor that the vents for this room were actually connected to the central air conditioning system. I was not welcomed with open arms by my colleagues when I arrived that morning.

In a fairly tense atmosphere, I removed the samples from the Berlese funnels and examined them for arthropods. In addition to the maggots I had collected from the corpse, I found larvae of moth flies, in the family Psychodidae, and larger soldier fly maggots. In decomposition studies in wet habitats I had previously seen larvae of psychodids on the corpse during the later part of the Post-Decay Stage. The maggots of the soldier fly were older than those I had collected from the corpse itself, the largest being 23 millimeters long.

There was a greater variety of beetles in the soil samples than on the corpse. In addition to adults of the scarab *Aphodius lividus,* which I had collected from the surface at the scene, there were also larvae in the soil. Although I did not recover any additional species of hide beetles from the soil samples, I did find more species of rove beetles. In addition to *Philonthus longicornis,* there were adults of *Philonthus discoides* and *Thyreocephalus albertisi.* And although I did not find any adults of the very large rove beetle *Creophilus maxillosus,* I did find larvae of this species along with larvae of a very small species in the genus *Oxytelus.* The presence of only larvae of these two species was again indicative of a postmortem interval of just over a month. The presence of only adults of the other three species was consistent with that time period, but not significant in delimiting it. There were also adults of two species of hister beetles in the soil samples, *Atholus rothkirchi* and *Saprinus lugens,* and some histerid larvae that I could not identify to the species level.

In addition to the insects in the soil samples, I found several other arthropods, including a specimen of a tailless whip scorpion. This animal is a very small, soil-dwelling predator, about 1 millimeter long. In Hawaii, it is most commonly associated with agricultural areas. I also identified a symphylid, an arthropod resembling a small centipede that feeds on plant roots instead of other animals. The samples also yielded several species of gamasid mites. Some were in the family Macrochelidae, whose members prey on fly eggs and young maggots. These were definitely related to the corpse, as were the Uropodidae mites, which feed on small nematode worms associated with decomposition.

In all, there were 23 different kinds of organisms associated with this corpse and the site of discovery. Not one of these had been actively involved throughout the entire decomposition process, but each of them had played a significant role in decomposition at a specific time.

I was faced with the problem of interpreting this arthropod evidence and putting it into perspective. First I selected decomposition studies whose environmental conditions most closely approximated the conditions in the pineapple field, combining data from two studies, one conducted inside Diamond Head Crater and one conducted on the campus of the University of Manoa. Both studies had environmental conditions and arthropod species in common. All of the species and stages of development I collected from the corpse and the pineapple field occurred at these two sites between days 30 and 40 of the studies. To limit this range, I looked at the developmental time required for the black soldier fly, *Hermetia illucens*, to reach the most mature stage collected from the scene and adjusted this according to the temperature data provided by the pineapple grower's weather stations in adjacent fields. This calculation narrowed my estimate to between 34 and 36 days before the discovery of the body on October 13, 1989.

There was some difficulty in identifying the corpse because of the advanced stage of decomposition. When an identification was made based on dental records, the victim proved to be a man who had been reported missing on September 15, 1989, and had last been seen alive on September 8, 1989. His blood-stained truck had been found in a parking lot on September 13, 1989. It had been 36 days between the last sighting of the man alive and the discovery of his body.

IN THE PREVIOUS two cases, I used the presence of species and stages of development to estimate the time of death. I have also encountered cases where the absence of life stages for a species proved to be a key piece of evidence. In one such case I was faced with the skeletal remains of a child approximately 30 months old, recovered from a grave on the side of Koko Head Crater on the island of Oahu on June 24, 1984. The child, a little girl, was buried on a small ledge overlooking Hanauma Bay and the

Pacific Ocean. The Honolulu Police Department personnel who recovered the body said that the skeletal remains were thinly covered with dirt and gravel, with some bones exposed and scattered on the surface. Four small stuffed dolls had been buried along with the body.

My examination of the remains took place on Monday, June 25, 1984, at the City and County Morgue. Because of the apparent absence of insects on the skeleton, the medical examiner did not at first think that I would be able to contribute much to the investigation, but he nevertheless contacted me and asked me to examine the remains.

When I saw the skeleton I immediately realized that the medical examiner had missed some very significant items in his assessment of the level of insect activity. Only a few types of insects

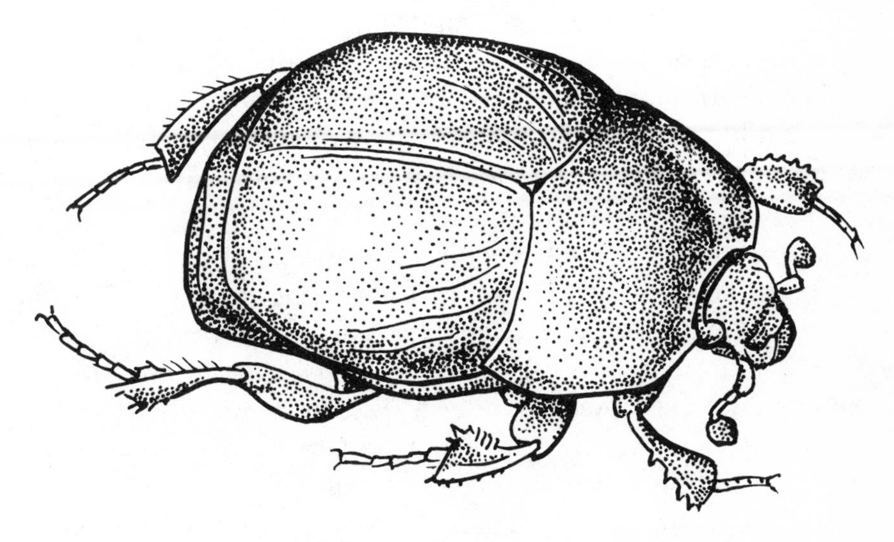

Larva (below) and adult (above) of the hister beetle *Saprinus lugens*

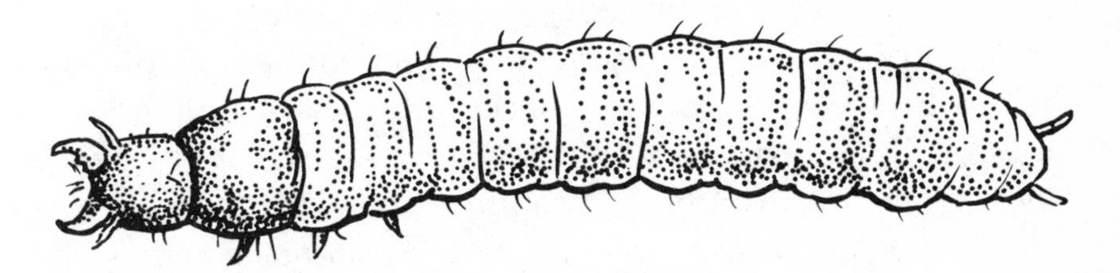

were present, but these still could provide some information about the death. The little girl had been buried in a pink hooded jacket and a pair of pink running shoes, clothing that assisted police in the initial presumptive identification of the victim. The hood of the jacket had been pulled close around the head when the remains were discovered, and it contained soil and by-products of decay from the skull. While examining the skull, I discovered under the remains of the scalp a number of empty pupal cases of the blow fly *Chrysomya rufifacies*, but no intact pupae. They indicated a postmortem interval of over 17 days. There were also adults and cast larval skins of the hide beetle *Dermestes maculatus* on the skull, but I could find no larvae or pupae of this beetle anywhere on the remains. Looking at the bases of the hairs on the scalp, I found larvae of a species of window fly, in the family Scenopinidae. In Hawaii, larvae of scenopinids are not commonly found on decomposing remains in dry conditions until 40 days or more after death. The specimens I collected were as large as those recovered from pigs on days 48 to 51 in studies I had conducted inside Diamond Head Crater. The child's right foot was still in a running shoe and was partially mummified. On this foot, I found an adult red-legged ham beetle, *Necrobia rufipes*, and an adult hister beetle, *Saprinus lugens*.

In addition to the specimens I was able to collect from the skeletal material, there were also the soil and decomposition by-products in the hood of the jacket to consider. I took this material and first examined it under a dissection microscope in my laboratory and then put it into a Berlese funnel for 48 hours to collect the microscopic specimens still in the soil. Under the dissection microscope I was able to discern additional empty larval skins of the dermestid beetles and some more scenopinid larvae. The Berlese funnel collection yielded more specimens of scenopinid larvae and a number of different species of mites. There were several predatory species of mites, including two in the family Macrochelidae, which feed on fly eggs and larvae as well as small soil-dwelling arthropods. There were two species of mites, *Macrocheles merdarius* and *Glyptholaspis americana*, represented only by adult females. I also found immatures and adults of species of mites in the family Uropodidae in the sample. These mites commonly feed on roundworms associated with decomposition. The soil under the corpse also contained the mites *Tyrophagus putrescentiae,* an unidentified species of the genus *Histiostoma,* and *Czenspinkia transversostriata,* all of which feed on by-products of decomposition.

My analyses of this assortment of species gave a preliminary estimate of 51 to 76 days between death and the recovery of the specimens from the remains. I then began to examine the patterns of activity and occurrence of stages of insect life cycles more closely. The larval scenopinids indicated a time period of from 48 to 51 days. The last larvae of *Dermestes maculatus* found in any decomposition studies conducted in similar habitats were collected 51 days after death. The presence of only adults of the two species of Macrochelidae was consistent with an interval of between 22 and 60 days. I estimated a total of 97 mites per milliliter for the sample from the hood of the jacket. According to decomposition studies this number was consistent with an interval of between 48 and 51 days, and the number of Acaridae in the samples indicated a period longer than 48 days. In the final analysis, the most significant factor proved to be the absence of larvae of *Dermestes maculatus* and the condition of the cast larval skins. In decomposition studies conducted in similar habitats, the last larvae had been observed 51 days after death. Once the larval skins are shed, they are quite fragile and degenerate rapidly if exposed to the elements on the soil surface. The cast skins I collected were in excellent condition and I could easily make a species-level identification. Their condition indicated a short period of exposure, and my estimate of the postmortem interval was slightly over 52 days.

In the meantime the police investigators had been busy dealing with conflicting statements about the crime. On May 3, the little girl's father reported that his daughter had been kidnapped by two men who arrived at his apartment on the night of May 2 and forced him and his daughter into a car. The father said they were then driven to the Chinatown area of Honolulu, where he was taken out of the car and beaten by one of the men. The two abductors then drove off with his daughter. The father was treated at a local hospital for injuries sustained in the beating. These events would have occurred 51 days before the discovery of the body. When the news media picked up this story, two men came forward and said that they had, in fact, beaten the father, but that the father had paid them to do so. They were usually willing to perform that service without compensation, but since they were being paid they did an exceptionally thorough job, thus landing the man in the hospital.

Confronted with this testimony, the father changed his story. He now said that he took his daughter hiking along the cliffs at Koko Head Crater on May 2, and while hiking he slipped and dropped the child over a 10-foot cliff. He climbed down to the ledge and found the girl still breathing, but unable to move. As he sat with her, her breathing and heartbeats became weaker. He could not revive her and she died. He then buried her in a shallow grave nearby. The next day, fearing prosecution, he reported the kidnapping. The police did not believe the accident story and issued a warrant for the father's arrest on a charge of murder.

Following his arrest, the father gave several different versions of the death, and by the time the case came to trial, the only aspect of it that had not changed significantly was my estimate of the time of the death. Both the prosecution and defense lawyers were happy with this estimate, because it suited both of their strategies.

In my first appearance in a criminal court, I had to convince a judge and jury that a significant aspect of a murder could be explained by looking at the empty skins of beetle larvae. Not only was this my first court appearance as an expert witness, but it was the first time entomological evidence had been introduced in a homicide trial in the state of Hawaii. In many respects, this was an ideal first case. Both sides wanted me to establish exactly the same time of death—the only time this has happened in all my courtroom experiences. Initially subpoenaed by the defense,

I was also consulted by the prosecution. During meetings with each side, I explained to both attorneys the basis for using insects and other arthropods as indicators of the postmortem interval, the techniques I had used in conducting decomposition studies, and how I had arrived at my conclusions in this case. After these meetings, I felt relatively confident, and arrived at court with few apprehensions. After all, both attorneys wanted the same basic testimony. But this confidence was soon shaken.

First I needed to be qualified as an expert in entomology. For some reason, just having a Ph.D. in the field was not sufficient. I was then still quite naive about courtroom procedures. All my education, professional experiences, and publications had to be presented to the court. I had supplied each attorney with a copy of my curriculum vitae, and essentially they took turns reading this to the court. No one, including the judge, had any idea of what constituted entomological competency. By the end of the first hour, I wasn't sure I was qualified to tell the time of day. Four hours later, it was decided that I was in fact a qualified expert in the field of entomology, and I began my testimony.

Unfortunately, as I began to testify, it dawned on me that neither attorney had any idea what I was talking about, except for the fact that insects were somehow involved and I had come to a conclusion with regard to the time of death. Then there was the unanticipated problem with the court recorder. Most of the words I was using were not in her dictionary, and I spent a good deal of my testimony time spelling scientific names for the record. I presented my results relatively quickly and cross examination was minimal and friendly because the lawyers each wanted my estimate to be believed. But even with an agreed-upon date of death, the father's version of events was not accepted by the jury. He was convicted of murder and is currently serving a life sentence.

6

Cover-ups and Concealments

In much of Hawaii the lava rock that constitutes the surface of a great part of the islands makes burial of a body difficult and time-consuming. But in other parts of the Hawaiian Islands and throughout many regions of the world burial is a common way of concealing a corpse.

The effects of even a shallow burial are remarkable. Each year during the portion of the course in Detection and Recovery of Human Remains that I teach at the FBI Academy in Quantico, Virginia, I place 50-pound dead pigs in different situations to demonstrate to FBI agents how bodies decompose differently under different conditions. The buried pig always presents the most dramatic contrast with the rest. During the course in May 1998, I had five pigs placed in different parts of a wooded area behind the FBI Academy. One was rolled inside a piece of carpet left open on both ends to allow insects to enter. A second pig was simply dumped on the ground. The third pig was hanging from a tree and not in contact with the ground, and the fourth pig was on the ground loosely covered by tree branches. The fifth pig was buried, but had only an inch of soil covering it. I put the pigs out at noon on the Thursday before the course began to allow for some decomposition before the students started the course on the next Monday. By Monday morning, all the pigs except the buried one, had very active insect populations. It was not until 7 days after burial that flies even began to be attracted to the buried pig. By this time all the other pigs had been reduced to skin and dried cartilage.

MY OWN EXPOSURE to delays caused by concealing or covering up a corpse began on the afternoon of December 31, 1988, with a telephone call from the medical examiner for the City and County of Honolulu. The Honolulu police believed they had discovered the body of a woman who had been missing from her home in Kahuku, on the north shore of Oahu, for approximately 2 weeks. They requested that I meet the medical examiner and follow her out to the scene, on the north shore of Oahu. At the time, my in-laws were visiting for the holidays and we were preparing for a dinner and a New Year's Eve party. So while the rest of the household dressed for a festive meal in town, I put on a selection of garments I reserve for such occasions—designed for comfort and not likely to show the stains that are inevitable in my line of work.

I must admit that a nice drive along the north shore on my motorcycle was not an altogether unpleasant prospect, and I arranged to meet the medical examiner near the Valley of the

Temples, a memorial park on the outskirts of Kaneohe town. Since the medical examiner was not familiar with the area, I led the way on my motorcycle. After an enjoyable 45-minute ride along the ocean to a spot just north of Amorient Aquafarms, we met an officer from the Honolulu Police Department near the intersection of an access road and the Kamehameha Highway. The access road led down to the ocean, running beside a golf course and some fields that eventually gave way to brush and scrub. The few houses around were well separated from each other and their views were blocked by the dense growth of koa haole trees, the most prevalent trees in the area. The crime scene was marked off with yellow tape and the beat officers and homicide investigators were waiting for us.

The bundle presumed to contain the corpse was well hidden in the undergrowth and might have remained undetected had the friends of the missing woman not been diligently searching for her. They had scoured this area because the woman's estranged husband was known to frequent it. When we arrived, the bundle had not yet been moved. One of the searchers had undone one corner of the wrappings and seen what appeared to be a human hand and wrist, but otherwise the bundle was intact. There were a lot of adult flies in the vicinity and on the outside of the blanket wrapping the corpse. Although no one had yet disturbed the corpse, there had been considerable activity in the immediate vicinity. Once the processing of the area around the corpse was complete and photographs were taken, I asked everyone to move away from the corpse to allow the flies to return and settle on the outside of the blanket. Since many species of flies produce larvae that are very difficult to identify as maggots but are relatively easy to identify as adults, I wanted to sample the population of adult flies around the corpse. These could provide clues as to the species involved before rearings to the adult stage could be completed. After about 15 minutes, I slowly approached the corpse and began sweeping my insect net over it and the nearby vegetation. After I finished my collecting and put the insects into containers, the wrapped corpse was removed to the side of the road, where it could be examined more easily.

Once the corpse was in clear view, we could see more details of the wrappings. The body was quite well concealed by a heavy brown blanket with the ends folded and tucked under. Both ends were firmly sealed shut with an Ace elastic bandage. Once this bandage was removed, with care taken to preserve the knots, the blanket was unwrapped. There was a second white blanket underneath. Moving through the folds of this inner blanket was a fairly large centipede. The police officer assisting with the unwrapping was not fond of centipedes and shortly there was not enough of the centipede left to bother preserving. As the inner blanket was removed, we saw that the head of the corpse was blackened and the abdomen was bloated. There were no immediately visible signs of trauma to the body. In a wet area like this one, as the body decomposes the outer layers of the skin separate from the underlying layers of tissue. When the body is moved, these outer layers slip away from the underlying layers of flesh. In this case there was obvious skin slippage, particularly in the arms and legs.

I found a large cluster of fly eggs on the outside of the outer blanket, along with a number of intact fly pupae. I could identify some of the latter immediately as being pupae of the blow fly *Chrysomya rufifacies* by their distinctive spines. Others appeared to be blow fly pupae, but of a different species. On the outside of the inner blanket, I found additional pupae of both the species I had already collected from the outside blanket. There were also blow fly maggots on the inner blanket. Predatory hister beetles, *Saprinus lugens,* were on the outside of this inner blanket and were feeding on the maggots, which had entered the post-feeding third instar stage and were attempting to leave the corpse. I collected representative samples from the body of all the maggots and numerous intact pupae, and I also took soil samples from the matted vegetation on which the corpse had lain, most of which appeared to have been alive when the wrapped corpse was deposited there.

I returned to my laboratory at the University of Hawaii with the samples. By the time I finished my initial processing of the specimens, it was 7:00 P.M. and the dinner reservations were for an hour later. I drove back across the island, showered, changed clothes, and then once again drove into Honolulu in time to join everyone for dinner and the New Year's Eve Party.

The autopsy was not conducted until the morning of January 3, 1989. I made additional collections from the corpse during the autopsy, but did not find any species or stages of development that I had not collected from the corpse at the scene. Four species of blow flies were present. *Chrysomya megacephala* was represented by second and third instar maggots and intact pupae. The maggots were restricted to the corpse itself, but the pupae were found on the corpse, between the layers of blankets, and on the outside of the blankets. I found third instar maggots of *Chrysomya rufifacies* on the same parts of the body as *Chrysomya megacephala,* and I observed more post-feeding third instars of *Chrysomya rufifacies* than there were of *Chrysomya megacephala.* I also collected two species of blow flies, *Phaenicia sericata* and *Phaenicia cuprina,* present only as third instar maggots. I had collected adults of all four species at the scene from the corpse and the surrounding vegetation.

As I anticipated, the soil samples from the ground under the corpse failed to yield any arthropods associated with decomposition. The corpse had been too well wrapped for such species to exit after feeding, and only plant-feeding species were present.

I found it significant that no empty pupal cases of any flies were present either on the corpse or in the area immediately around it. This suggested that I was still working with pupae that had developed from the first eggs laid on the corpse and that the time required for development from egg to adult would closely approximate the time elapsed since the onset of insect activity on the corpse. I closely monitored the pupae I was rearing in the laboratory and noted the first emergence of an adult *Chrysomya megacephala* at 3:00 A.M. on January 2, 1989. This emergence was followed 48 hours later by the emergence of nine more adults of that species and the first adult of *Chrysomya rufifacies.* No adults of either species of *Phaenicia* emerged from their pupae. Using data obtained from the rearing of *Chrysomya megacephala* under controlled conditions in my laboratory, I determined that 6,415 ADH were required for *Chrysomya megacephala* to complete development from egg to adult. In this case, 858 ADH had elapsed while the pupae were inside the environmental chamber in the laboratory at 26°C. This left 5,557 ADH that would have elapsed while the flies were on the corpse.

Adjusting for ambient temperatures at the scene was fairly simple because there was a NOAA weather station (ironically station number 911) at Amorient Aquafarms, about half a mile from the crime scene. I compared temperatures between the scene and the weather station and found no significant variations. Using the weather data from that station, I determined that the 5,557 ADH required to complete development at the site was consistent with an onset of insect activity 10.5 days before the discovery of the body.

Through interviews with witnesses, investigators determined that the woman had last been seen alive 13 days before the corpse was found. At that time she was in her home in Kahuku with her estranged husband. That afternoon, a neighbor heard a gasping sound and a thumping, as if someone was being pushed against a chair. Later that day the woman was seen with her husband in his pickup truck, sitting motionless in the passenger seat with her head leaning back. My estimate left a gap of 2.5 days that was unaccounted for by the insect activity. I reasoned that the wrapping of the corpse had delayed the flies' access to the body and thus the onset of insect activity. In my report to the medical examiner, I gave my estimate of 10.5 days of insect activity and also noted the probability that the onset of insect activity had been delayed by the wrapping of the corpse in two layers of blankets.

My report was acceptable to the medical examiner, the Honolulu Police Department, and the city prosecutor in charge of the case. The victim's estranged husband had been charged with the murder and the trial was scheduled. At the trial, the defense attorney presented a scenario in which somebody else killed the woman during the 2.5 days that could not be accounted for by insect activity. The defense attorney interviewed me before the trial, questioning me closely about this gap, and asking if there was any way my estimate could be more accurate. Since he continued to press me about this in subsequent telephone calls, I thought I should investigate further by performing a relatively simple experiment.

I obtained a 50-pound pig from a commercial slaughterhouse on the windward side of Oahu and duplicated the wrappings on the corpse. I then looked for an area similar to where the corpse had lain, and found it in my own overgrown backyard. I put the pig in an exclosure cage to see how long it would take flies to penetrate the wrappings and lay eggs on the pig. I examined the pig every 4 hours, carefully preventing flies from accessing the pig during sampling by covering myself and the pig with a small mesh net.

Several things happened during the course of this experiment. I had considerable support from my dog and my daughter's cat, who seemed to think this was the only useful thing I had done in years. I discovered that my wife and daughters were even more tolerant than I had expected. And I was surprised to learn how much attention a couple of my neighbors paid to what went on in my backyard. Several years later, this case was re-created for a television program. The same neighbors, again peering from behind curtains, were treated to the spectacle of my repeatedly wrapping and unwrapping another dead pig so that we could be filmed from every possible angle.

From the original experiment, I learned that it took 2.5 days for the adult flies to penetrate the wrappings and lay eggs on the pig. Combined with the period of insect development, this period accounted for the entire 13 days since the woman had last been seen alive.

AS THE PREVIOUS case shows, a relatively complete covering of a corpse significantly delays the onset of insect activity. But a loose wrapping does not appear to have much delaying effect. In one such case, a clothed corpse was found wrapped in a canvas car cover near a popular stream beside and below the Old Pali Road in Honolulu. The victim, a male, had been shot and his body dumped over the side of the road. The corpse was discovered by a man walking with his son along the shoulder of the road on June 3, 1996. He had noticed the pile of canvas on a previous walk but thought little of it. This time, flies were buzzing around the canvas and the unmistakable smell of decomposition was rising to the road above.

I was called to the site by the medical examiner and arrived there about 11:30 A.M. The corpse was down a steep incline below the road, and an emergency crew from the Honolulu Fire Department was lowering a ladder for access. Firefighters recovered the corpse and brought it, still wrapped, to the road. Once the crime scene technicians had taken their photographs and samples from the scene, I went down the ladder to take soil and vegetation samples. The local news media had arrived early in the process, and although they were kept some distance from the crime scene, and were prevented from getting a good shot of the wrapped corpse being brought to the top, they were able to document my trips up and down the ladder. Since the victim and all the suspects were Marines, ultimately this case came under military jurisdiction, and military police allow few photo opportunities. So each time the case was mentioned on one local television station, file footage of my trip up and down the ladder was shown.

As the corpse was unwrapped at the scene, I noticed that there was tape over the mouth and nose and what appeared to be a wound to the back of the head. There were a number of different species and stages of insects on the body. The wrapping had not been sufficient to significantly delay the invasion of insects. Four different families of flies were on the corpse: pupae and empty pupal cases of *Chrysomya rufifacies;* third instar maggots of a relative of the house fly in the genus *Fannia;* larvae of the cheese skipper *Piophila casei;* and larvae of the black soldier fly, *Hermetia illucens.* There were numerous larvae of *Hermetia illucens* on the canvas wrapping material, as well as three families of beetles on the outer surfaces of the corpse and on the canvas. I collected adults of the red-shouldered ham beetle, *Necrobia ruficollis.* I found adults of the hide beetle *Dermestes ater* on the corpse, but not any larvae. The rove beetles were represented by adults of two species commonly associated with decomposing bodies: *Thyreocephalus albertisi* and the large *Creophilus maxillosus.* I put all these specimens into rearing containers for transport back to the laboratory.

In the laboratory, I fixed and preserved the beetle specimens collected from the corpse and processed the fly larvae into sublots as usual. The soil samples were processed in a Berlese funnel for 48 hours to extract the soil-dwelling organisms. Since the disposal site was on a steep incline thickly covered by vines, there was only a thin layer of soil, offering few habitats for insects. This scanty soil cover combined with the wrapping of the corpse limited the specimens I collected to some *Hermetia illucens* larvae and a few empty pupal cases of *Chrysomya rufifacies.*

During the autopsy at the City and County Morgue on June 4, I collected more intact pupae of *Chrysomya rufifacies* and larvae of *Hermetia illucens* similar in size to those I had collected from the corpse the day before. I also found additional specimens of the cheese skipper *Piophila casei,* as both larvae and pupae, in the corpse's clothing. From the body and clothing, I also collected earwigs in the family Labiidae, adults of the red-collared ham beetle, *Necrobia ruficollis,* and adults of the rove beetle *Thyreocephalus albertisi.* In addition to the adults of the hide beetle *Dermestes ater* that I had observed at the scene, I found early and mid instar larvae while examining the corpse at the morgue. Another addition to the fauna was a species of rove beetle, *Philonthus longicornis.* Comparison of the different species and life stages I collected at the morgue and at the scene shows the need for collections at both places. Species and developmental states that I might miss in the field collections I can frequently find during more controlled examinations at the morgue. And some species and life stages I can collect at the scene may leave the body during placement of the corpse into the body bag and transport to the morgue. If I am not able to make both examinations, there is a very good chance I will miss significant evidence.

The specimens I collected in this case were a mixed assortment. The habitat where the body was found was quite similar to a rain forest, but there was enough sunlight to allow the body and the wrappings to dry out slightly. Some hide beetles were present on the body but if the corpse had been lower in the ravine, it probably would have been too wet for these beetles to have colonized it. The relatively heavy rainfall in the area, combined with the water retained by the canvas wrapping, had kept the body moist and pliable longer than usual, allowing more than one generation of blow flies to develop on the corpse. The species in the genus *Fannia* are typically associated with wet areas in Hawaii and are characteristically found after several weeks of decomposition. The *Fannia* larvae combined with the cheese skipper pupae indicated a time frame of slightly over a month. So did the size of the larvae of the black soldier fly, *Hermetia illucens,* a species not normally attracted to a decomposing body in Hawaii until approximately 20 days after death. The larvae collected from the corpse and soil samples would have required approximately 9 to 11 days to reach their stage of development. All these things considered, the estimated postmortem interval was 29 to 31 days before June 3, 1996.

There were several suspects in this case and their confessions ultimately placed the time of death approximately 30 days before discovery of the corpse. The victim had been lured to the suspects' apartment on the pretense of resolving a debt. According to confessions and subsequent reported testimony, the suspects had initially planned only to beat up the victim, but the encounter escalated into an execution-style murder.

ALTHOUGH WRAPPING AND burial are the ways of attempting to conceal a corpse I have most frequently encountered, other efforts to cover up have proved challenging. One of the earliest of these cases involved the body of a woman found on July 26, 1983. The bloated corpse was discovered when military police responded to an anonymous telephone call reporting a foul odor coming from an apartment in military housing in Honolulu. While checking the apartment, a military police unit opened a closet door and found a woman's body lying on its back on top of a pile of laundry. She was wearing a red floral dress that had been pulled above her navel and a pair of blue bikini panties. The head was blackened and decomposed.

I examined the body at the morgue on the morning of July 27, 1983, and collected specimens of two species of fly maggots and pupae. The natural body openings of the head, anus, and genitals were infested with maggots. I also saw maggots on the trunk and limbs, but these appeared to have migrated away from the main sites of infestation, probably because of the disturbance when the police moved the body. I collected pupae from the legs. There was an exceptionally large mass of maggots associated with the genital openings, possibly related to the pregnancy of approximately 6 to 7 months duration, which was discovered during the autopsy. But the maggots had not invaded the major body cavities.

I divided the maggots and pupae into two sublots and processed them as usual. I identified blow fly maggots as *Chrysomya megacephala* both from the adults emerging from the pupae collected from the body and from maggots I reared in the laboratory. The other species of fly was *Synthesiomyia nudiseta,* a species commonly associated with corpses discovered indoors. The stages of development collected from the corpse combined with total developmental times for both species of flies gave an estimated postmortem interval of approximately 8 days.

The suspect in this case, the victim's husband, initially admitted to having strangled his wife at approximately 3:00 P.M. on the day before the discovery of the body. He made this confession while the investigators were still at the scene, and when they called the suspect's attention to the obvious state of decomposition of the corpse, he changed the time frame to "about a week ago."

In this case, there was never a precisely determined time of death; some of the information gathered from interviews with neighbors tended to support the interval given by my analyses of insect development. This couple had had heated domestic arguments on a routine basis, and since these arguments were usually quite loud, the military police had frequently been called to the apartment, sometimes two to three times a week. The 8-day period covered by my entomological estimate was considered remarkable by the neighbors for its lack of domestic disputes and police appearances. The victim was last seen alive by neighbors on July 18, 9 days before discovery of the corpse, and the last domestic dispute anyone could recall was on July 19. From the available evidence, it appears that the husband strangled his wife on July 19 and immediately put the body in the closet. He then continued to live in the apartment along with his 20-month-old daughter for the next 8 days, until neighbors reported the odor. A known substance abuser, the husband was never able to recall more of the event than that he had strangled his wife "a little over a week ago."

A MORE BIZARRE method of concealment was used in the case of an 11-year-old girl reported missing from a town in northern California. She had last been seen alive at approximately 1:45 P.M. on February 25, 1996, and was reported missing by her parents at 5:00 P.M. the same day. When last seen she had been on her way to visit a friend who lived only a couple of blocks away. The friend denied any knowledge of her whereabouts. A search was initiated by local law enforcement personnel with the assistance of the FBI and volunteers. On March 21, 1996, the friend's mother smelled a strong, unpleasant odor and asked her son to find what was causing it. At approximately 7:00 P.M., the son called the local police and reported that he had smelled the odor of something dead coming from a ceramic kiln located near the side of their house. He had opened the lid of the kiln and found what was later identified as the body of the missing girl.

The kiln was an electric ceramic kiln with a tight-fitting top and three vents on the sides, each approximately an inch in diameter. Two of these vents were open, but the third was blocked. The kiln had not been used for 5 years. When the body was removed from the kiln, it was found to be wrapped in a sheet and the girl's coat. The skull and chest were largely stripped of flesh to the level of the diaphragm, with the lower portions of the body remaining intact. The medical examiner had previously attended workshops I had conducted, along with another entomologist, a pathologist, and forensic anthropologist, for the American Academy of Forensic Sciences. During the autopsy, which was conducted on March 22, 1996, he took samples of the maggots, following the guidelines we had laid out for the collection and preservation of maggots.

I was contacted by an investigator for the district attorney's office and agreed to examine the specimens. One batch of the maggots had been kept alive, with beef liver to feed on. These were being watched by an investigator for the local police department—not the most pleasant task under the best of circumstances. The other sample had been fixed and preserved in ethyl alcohol. The specimens were shipped to me via Federal Express and arrived in my laboratory on the afternoon of April 10, 1996.

I put the live specimens in the environmental chamber at 26°C for development to the adult stage and began identifying the preserved maggots. There appeared to be two species of blow flies in the sample and a single third instar maggot that appeared to be in the family Muscidae. I identified the blow fly maggots as *Phormia regina* and a species of *Phaenicia.* The single muscid maggot had quite distinctive posterior spiracles and I was able to identify it as *Synthesiomyia nudiseta.* The live specimens began to pupate on April 11 and adults began emerging on April 14. The adults confirmed my identification of *Phormia regina,* and I was able to identify the other species of blow fly as *Phaenicia sericata.*

On April 12, I received another shipment from the district attorney's investigator. This one contained the weather data and scene photographs I had requested earlier, as well as a videotape of the body being removed from the kiln and the collection of maggots during the autopsy.

I had suspected that the time required for the maggots in the sample to develop would not be anywhere close to the period of time between the last sighting of the child and the discovery of the corpse. Once the identifications were confirmed, it appeared that I was right. The most mature maggots were post-feeding third instars of *Phormia regina.* I analyzed the temperature data from the weather station for the period from February 24 to March 22, 1996. The time necessary for these maggots to reach this stage of development indicated that the eggs were laid on the corpse between March 14 and 15, 1996. The child had been reported missing long before that, on February 25, 1996.

I then looked for more temperature data on blow flies to see if I could find any additional information that might shed light on the case. I found that during his doctoral research in Indiana in 1990 Neal Haskell during his doctoral research had determined that most blow flies in the continental United States, including *Phormia regina,* are not active at temperatures below 12.5°C. Digging further, I found that Allan Pfuntner had also investigated egg laying by *Phormia regina* while doing research for his master's degree in San Jose, California, in 1977. He had found that *Phormia regina* do not lay eggs at temperatures below 20°C.

With this information, the situation became a little clearer. The weather data showed that between February 24 and March 7, there were few times that temperatures were high enough for the flies to be active and it never got warm enough for them to lay eggs. Between March 7 and March 10, there were a few periods of 2 to 6 hours' duration when temperatures rose high enough for egg laying, but it is doubtful that the flies would have had enough time during these brief periods to locate a corpse hidden in a kiln with only two small openings for access. After March 10, temperatures once again fell below the threshold required for egg laying and remained there until March 15. From March 15 until the corpse was discovered, temperatures were well above the threshold required for egg laying.

In this case, I was not able to establish a postmortem interval that corresponded to the total time between the disappearance of the child and the discovery of the corpse. There were several factors in addition to the low temperatures that limited the activity of the flies: limited access to the corpse for egg laying; the additional time required for the flies to locate the corpse; and the fact that the interior of the kiln was tightly sealed off except for the two open vents and almost completely dark. Flies will enter a closed, darkened place to lay eggs, but they must first be attracted by strong odors of decomposition.

Given these facts, I concluded that the child had been put inside the kiln shortly after her death. The temperatures during the first 11 days after February 25, when the girl had been reported missing, remained too low for the flies to lay eggs. Then came a period of temperatures high enough for egg laying, but too brief for the flies to locate the body, followed by a short period when the temperature again fell below the threshold. When the temperatures again rose to the level at which egg laying could occur, the corpse had decomposed sufficiently to become highly attractive to the flies, and the first eggs were laid, most probably on March 15. The most mature maggots had developed into the post-feeding third instars I received. Although the insect activity could not account for the entire period, those specimens and stages of development I examined were consistent with the facts discovered about the case, although no one was ever arrested for the murder.

OTHER ATTEMPTS WERE made to conceal the body in the case of the girlfriend of the John Miranda, who on February 6, 1996, returned to his former place of employment in Honolulu and took a former co-worker hostage. The drama played itself out when he emerged with his hostage and began to descend a flight of stairs on the side of the building. As they reached the bottom, the hostage suddenly twisted away and John Miranda was killed by a single shot from a police sharpshooter. This sensational event was recorded and replayed for several days by the local news media and later was incorporated by a mainland production company into a documentary on hostage situations.

Miranda's death resolved the hostage situation, but did nothing to clarify the fate of his girlfriend, Sherry Lynn Holmes. A hostess at a bar in the Moiliili area of Honolulu, she had not been seen alive since January 31, 1996. After his death, it was learned that John Miranda had reportedly told friends that he had killed and buried her. The search for Sherry Lynn Holmes centered originally on the windward side of the island of Oahu, along the Kapaa Quarry Road—a popular site for disposal of human remains because it is heavily overgrown and already littered with illegally dumped rubbish. A wide-ranging search yielded nothing.

Then, on March 31, 1996, acting on a tip from an informant, the police returned to Kapaa Quarry Road with a cadaver dog and located the grave of Sherry Lynn Holmes. It was about 40 to 50 feet from the road and approximately 3 feet deep. The body was in a cardboard box secured with rope and duct tape, and the head was covered with a plastic bag. Had John Miranda killed her and buried her body shortly before he himself died in the hostage incident, or had she been killed by somebody else after his death?

Since the grave was shallow, insects and other animals had reached the corpse without much effort. As the body was removed from the grave, I saw several different kinds of animals associated with the corpse and the cardboard box. Numerous earthworms were in the soil and moving through the box. I collected several specimens of millipedes in the genus *Spirobolellus* from the box, along with the larvae of a clothes moth (family Tineidae). I found empty pupal cases of both species of blow flies commonly found with decomposing remains, *Chrysomya megacephala* and *Chrysomya rufifacies,* on the box and the clothing. From the body itself, I also collected third instar larvae of *Chrysomya megacephala.* I realized that these would not normally be present on a body in the state of decomposition indicated by the rest of the organisms, and were most probably related to the moist earth at the bottom of the grave. This wetness had kept the tissues pliable and suitable for development of maggots far longer than a dry habitat would have. This assumption would be supported by the absence of any species of hide beetles in the family Dermestidae, since the corpse must dry before they can adequately exploit it as a food source. In this case, the corpse had not dried out enough for them, even though there were dermestids in the area on the drier surface of the soil.

Many other insects were present. I noted adults of the muscid fly *Ophyra aenescens* emerging from their pupal cases while I was examining the corpse and the clothing. In the clothing, I found adults and larvae of two species of rove beetles, *Creophilus maxillosus* and *Philonthus rectangularis,* and in the soil of the grave I found adults and larvae of two other species, *Philonthus discoides* and *Thyreocephalus albertisi.* I also found larvae of the red-legged ham

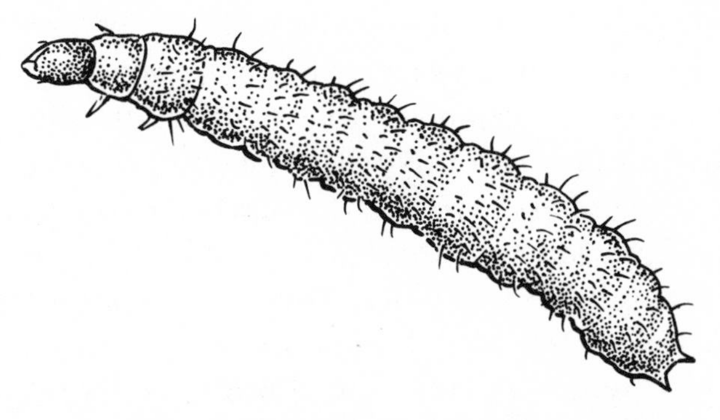

Larva of the red-legged ham beetle, *Necrobia rufipes*

beetle, *Necrobia rufipes,* in the soil, but none on the corpse itself, and no adults anywhere. Significantly, on the corpse and clothing, and in the soil of the grave, were larvae of the black soldier fly, *Hermetia illucens.* The most mature of these larvae were 24 to 26 millimeters long. There were also three more families of beetles represented in the soil I collected from the grave and processed with a Berlese funnel: Histeridae (hister beetles, *Bacanius atomarius*), Nitidulidae (sap beetles), and Tenebrionidae (darkling beetles). All these were adults; I saw no larvae.

I used climatic data from the NOAA weather station closest to the site in my calculations. The main insect I used to estimate the postmortem interval was the black soldier fly, *Hermetia illucens.* If the remains are buried, this species is not attracted for at least 30 days under Hawaiian conditions. Once the flies lay eggs under the conditions in which this corpse was found, a minimum period of 26 days would be required for the development of larvae as large as those I collected from this corpse. Thus 56 days was the minimum postmortem interval. The remains were discovered 54 days after the death of John Miranda, so he could have killed Sherry Lynn Holmes before he was shot, as informants reported he had told them.

SOMETIMES THE MURDERER'S attempts to cover up the crime have unintended consequences that significantly affect efforts to estimate the postmortem interval. I was introduced to one such case late one evening in October 1994, when I was called to look at a corpse found just off the road leading to Round Top State Wayside Park above Honolulu. The body was about 15 feet below the level of the road, lying on a grassy slope. It was fully exposed, with nothing except the clothing—a tan-colored tank top and cut-off jeans—blocking it from view, keeping the sun off it, or otherwise inhibiting insect access. On trees only a few feet from the body someone had posted fliers asking for help in locating a missing Makiki man.

It was still light when the corpse was discovered, but night was falling when I arrived at the scene and it was completely dark by the time I began to collect specimens. The slope was so steep that the corpse had to be was brought up to the road in a rescue basket deployed by the Honolulu Fire Department. The vegetation was so dense under the corpse that I did not think I would recover any arthropods from a soil sample that would be useful, but I took a sample anyway. I was right, however; after processing the sample in my laboratory, I found several cockroaches, some springtails (order Collembola), a couple of terrestrial sand hoppers, and a few soil-dwelling mites, but no species definitely related to decomposition.

From the corpse, I collected fly egg masses and a number of maggots, some of which appeared to be larvae of two species of blow flies. Even though it was well after dark at the time, adult flies were still laying eggs on the corpse and they continued to do so while I made my collections. Obviously, the flies were not aware that they were supposed to stop laying eggs as soon as darkness fell. As usual, the samples were fixed, and representative specimens were put into the environmental chamber at 26°C to be reared to the adult stage. The next morning, I went to the morgue to take additional samples from the corpse during the autopsy. I collected maggots from the outside of the body, from natural body openings on the head, and from the visceral cavity. These last maggots had gained access to the abdominal cavity through wounds. I treated these specimens the same way as those I had collected the night before.

I identified the maggots as blow flies of *Chrysomya megacephala* and *Chrysomya rufifacies*, and a third species in the fly family Muscidae. There were second and third instars of both *Chrysomya* species, but the Muscidae family was represented only by a single specimen, a third instar maggot that appeared to be the common house fly, *Musca domestica*. The weather data needed for the calculation of the time required for development of the two species of blow flies I obtained from the weather station at the Lyon Arboretum, about 2 miles from the scene. After adjusting the data with a linear regression, I was able to estimate that there had been approximately 5 days of insect activity on the corpse before I collected my samples on the evening of October 12.

When the victim was identified, he proved to be the missing Makiki man described on the flier posted at the scene. He had been missing for 7 days before his body was discovered. Since insect activity accounted for only 5 days, I had 2 days left to explain. Had the corpse been covered or concealed in some manner, I could easily have understood this gap. But the corpse had been completely exposed, and I should have been collecting prepupal or post-feeding third instar maggots and no second instar maggots of either species of blow fly. Instead, I had a collection of blow fly maggots typical of a 5-day postmortem interval. There was, however, one species that was consistent with a 7-day interval, the house fly, *Musca domestica,* represented by a single maggot. In my decomposition studies, I have found that the house flies and their relatives do not arrive at a carcass until a couple of days after death, and the fact this maggot was in the third instar fit a period of 7 days, not 5. The evidence seemed to conflict.

Unknown to me—and everyone else—for some time after the discovery of the body, the body had been moved, and there was more than one crime scene involved in this murder. These two scenes were not connected until days after I had completed my preliminary estimate. Shortly after the victim had been reported missing, the police had been called to a parking garage in downtown Honolulu, where they found an enormous amount of blood but no body. The police and the investigators from the medical examiner's office were both convinced that no one could possibly have lost that much blood and survived without emergency medical attention. But the police could find no record of anyone seeking medical attention that night who had lost significant quantities of blood.

Several days after the discovery of the corpse on Round Top

Drive, the two scenes were connected through accounts of the victim's activities, and some reasons for the apparent lack of insect activity emerged. The victim had last been seen alive at approximately 1:30 A.M. on October 5, 7 days before the discovery of the corpse, at which time he had been on his way to meet another man to collect a gambling debt. It was later shown that he had been killed in the parking garage that night and that the fatal wounds were so severe that his body was virtually drained of blood at the scene. The murderer then took the body to Round Top Drive and threw it over the side of the road. By the time the body was dumped, almost all the blood had drained not only from the wounds but also from the mucous membranes of the eyes, nose, and mouth, leaving little to attract adult flies, except house flies and their relatives, which respond less to blood than to decompositional fluids. The wounds did provide an immediate means of entry for maggots, which were disturbed by the removal of the body from the hillside. But, I thought, adult flies would not have been attracted until decomposition by anaerobic bacteria in the digestive system had reached the point where fluids began to ooze from the body openings. Then and only then would adult flies arrive and begin laying eggs at the natural body openings. Thus the stages of development for all three species of flies I collected made sense. When the killer, who was later convicted of murder, was arrested, his version of events confirmed my timetable.

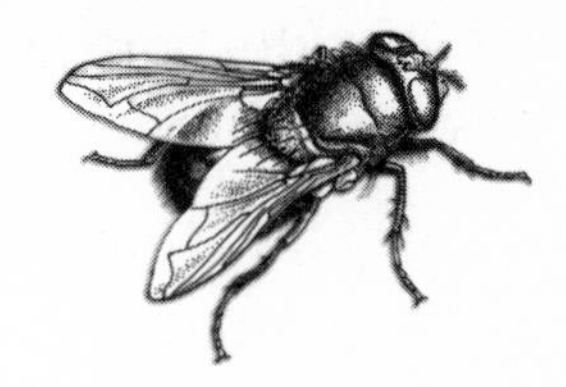

7

Predators

Predatory arthropods are a major component of the fauna I encounter at death scenes. Usually—when these predators are attracted to the corpse by the activities of the necrophagous species feeding on its tissues—they have a predictable place in the succession patterns and do not unduly alter the rate at which the corpse is consumed. But one group of these predators, the social insects, can in some circumstances significantly alter the rate of reduction of the corpse.

The social insects—ants, wasps, bees, and their relatives—form colonies in which different groups of individuals, called castes, are specialized to perform particular functions. With each caste doing its job, the colony as a whole becomes very powerful, and the castes specialized to obtain food are very efficient at doing so because other members of the colony take on the tasks of reproduction, raising the young, and maintaining and protecting the colony.

WHEN A CORPSE is deposited near a colony of social insects, the results can be dramatic. During the 1994 meetings of the American Academy of Forensic Sciences in San Antonio, Jack Hayes described a case where imported fire ants, *Solenopsis invicta,* killed and removed the necrophagous species as soon as they arrived at the corpse. So severe were these ants' stings that the people attempting to recover the body had great difficulty in doing so.

I have observed similar ant behavior during my studies in Hawaii. In one of my early experiments inside Diamond Head Crater, I unknowingly placed one of my pig carcasses close to a colony of another species of ant, *Solenopsis geminata.* I saw only a few ants during the first couple of days of decomposition, but by the fourth day there was a dramatic increase in the ant population. Arriving early that morning and looking at the site from a distance, I saw what appeared to be white chalk lines someone had drawn in the soil, leading away from the carcass. As I got closer, I could see that they were not chalk lines, but columns of ants. Each ant was carrying a white maggot away from the carcass and back to the colony. These trails were eventually worn over a quarter of an inch deep in the soil. Later, the ants established secondary colonies right next to the tail of the pig and, still later, in other areas of maggot activity. All in all, the ants carried away so many maggots that consumption of the carcass by the necrophagous species was significantly slowed—by several days!

SOME WASPS CAN also be fierce predators of necrophagous species. While working on the island of Hawaii on a study comparing altitudinal variations in rates of decomposition and the composition of arthropod communities associated with decomposing remains, one of my graduate students observed a fierce attack on flies and maggots by one of the native Hawaiian wasps. This species, *Ectimneus polynesialis,* in the family Sphecidae, is known only from the Hawaiian Islands, and was preying on flies at a site in an upland forest habitat 1,870 meters above sea level. These wasps were so effective at capturing and removing flies from the carcass that they extended the normal period for fly egg laying on the carcass.

In addition to this species, I have also observed an introduced (nonnative) wasp in the family Vespidae, *Vespula pennsylvanica,* on both Oahu and Hawaii at a wide range of elevations preying actively on adult flies and maggots. Whenever I see any species of social insects associated with a corpse, I have to consider whether their depredations have altered the normal pattern of decomposition. •

THE SIGNIFICANT REDUCTION of the population of necrophagous species by social insects can in itself, under certain conditions, provide valuable clues to the history of a corpse. This happened in one of the strangest cases I have ever encountered. The story begins in the early evening of October 24, 1994. A carpenter on his way home from his job was traveling along the old Kalanianaole Highway on the windward side of the island of

Oahu. In his spare time, he had been restoring a pickup truck, and all he now needed was a toolbox for the bed. As he was driving, he spotted what appeared to be a toolbox along the side of the road. When he opened the box to see if he was in luck, he found not tools, but a human skeleton.

The crime scene in this case was limited to the metal toolbox—the only completely portable crime scene I have ever seen. The box was easy to pick up and transport to the City and County Morgue for a controlled examination, and I first saw it there on the following morning. Inside the box were the skeletal remains, some fragments of cement, a few white plastic trash bags, a "one-size-fits-all" T-shirt, and some dirt. The skeleton wore a short-sleeved aloha shirt, briefs, and short jeans. The feet had socks and steel-toed construction boots in place. Also on the skeleton were a watch and a digital pager.

The first thing I noticed on entering the room where the investigators were processing the box was that they were all brushing themselves off. This is unusual in a morgue, where people usually try to avoid contact with, and thus possible contamination from, any organisms that are present. As I got closer to the box, I saw the problem—ants. They were everywhere, but the bulk were coming out of the toolbox—so many ants that processing the skeleton was almost impossible. Even the overnight refrigeration of the box had not appreciably slowed the ants' activity, and as the box began to warm, they got more active. Everyone in the room was very eager for me to make my collections and give my OK for a dose of ant-and-roach spray.

The assortment of insects on the skeleton and in the toolbox was quite unusual. In the dirt and debris at the bottom of the box, I found empty pupal cases of the blow fly *Chrysomya megacephala* and of the black soldier fly, *Hermetia illucens*. There was also an empty egg case of a cockroach on the side of the box. I collected adults of the scarab beetle *Onthophagus incensus* and adults and immatures of the earwig *Euborellia annulipes*. Actively moving on the inside and outside of the box were numerous adult workers of the long-legged ant, *Anoplolepis longipes*, and there were also several larvae of that species in the debris. The numbers of adult ants and the presence of larvae indicated that the colony was somehow associated with the box, but I did not find a colony in the debris.

Everybody there could see that the skull was the center of ant activity. I looked closely at it and discovered that the cranial vault housed not the brain but an ant colony. In addition to the numerous adults, the skull contained larvae and pupae. I collected representative samples of all stages and forms present. There were two distinct sizes and shapes of larvae in the colony and also two sizes of pupae. Even though most ants were on and in the skull, the disturbance caused by the moving of the skeleton and the box had clearly caused panic in the colony, and workers were carrying larvae and pupae out of the skull as I was collecting my samples.

In addition to the ants, I also recovered a number of empty pupae of the black soldier fly and of the blow flies attached to the various bones of the arms and legs and on the ribs.

After I had completed my collections, a can of ant-and-roach spray materialized as if by magic and the contents were emptied onto the skeleton and the box. Once the insects were officially dead—actually overkilled—the physical anthropologist and odontologist were able to begin their examinations. As is typical with many sets of skeletal remains in Hawaii, determination of race was somewhat problematic. The physical anthropologist determined that the victim was a male "Mongoloid, with possible Caucasoid admixture," about 50 to 60 years old. There was a good deal of dental work that, according to the odontologist, would have been done over a considerable period of time and would have cost a lot of money. Often such extensive dental work helps with identification, but in this case no matching dental records could be found—a fairly common circumstance among the very mobile population of Hawaii. Some clues to identity had been obtained from the serial number on the digital pager, and a presumptive identification was made by superimposing videos of the skull on photographs of the owner of the pager.

Meanwhile, I was examining the insect evidence. This was a situation I had not encountered before. I frequently find adult ants and wasps preying on maggots that are feeding on a corpse, but because of the transient nature of a decomposing body, I almost never see whole colonies of social insects on a corpse. I was not sure initially how this ant colony's presence might affect my ability to estimate the postmortem interval.

So I first turned to those insects I was more accustomed to finding on a corpse, the flies. Both blow flies and black soldier flies were represented on the skeleton and in the debris from the box by empty pupal cases. The presence of pupal cases of the blow fly *Chrysomya megacephala* indicated a minimum period of at least 17 days, the usual time required for development from egg to adult. Since the body had been concealed in a box, the adult flies could not have found the corpse immediately, and the time required for development of all the eggs to adults would thus be somewhat longer than 17 days. Furthermore, because the body had been reduced to a skeleton, it was obvious that 17 days plus some days added for the delay in the adult flies' arrival was still far short of the time since death.

The presence of only empty pupal cases of the black soldier fly indicated a longer period of decomposition because under Hawaiian conditions this species typically does not invade a corpse for 20 to 30 days after death, depending on the circumstances. In this case the enclosure of the body in the toolbox most probably indicated a period of 30 days or so. The time required for this species to complete development from egg to adult is somewhat variable, ranging from 5 to 7 months, depending on temperatures. In Hawaii, the developmental period is approximately 5 months under most conditions. Together these two times indicated a minimum period of about 6 months of insect activity that I could document.

Aside from the ants, the remainder of the insects collected from the box and the skeleton were species incidental to decomposition and did not provide any additional clues as to the postmortem interval. I therefore turned my attention to the ant colony. This species of ant, *Anoplolepsis longipes,* is widely distributed in Hawaii, usually found in fairly dry areas, and is prone to form colonies under rocks. This colony, though large enough to cause considerable discomfort in the confines of the morgue, was not very big by species standards, and I thought it was quite a young colony. In Hawaii ants are considered serious pests because they associate with various species of mealybugs that are quite damaging to agricultural products, particularly pineapples. The ants protect and maintain the mealybugs on the plants, receiving in return the honeydew—sweet and nutritious excrement—secreted by the mealybugs. This interaction between ants and mealybugs has led to a number of studies of ant biology and behavior undertaken to assist in control efforts. I consulted with Jack Beardsley, who along with his graduate students has spent a number of years investigating ant and mealybug problems in Hawaii. He agreed with my assumption that this was a relatively young colony.

This conclusion gave me a time frame to work with. In a newly established colony, the queen first must produce the workers that will rear the immatures, defend the colony, and gather food. Typically, about a year elapses before any additional reproductive forms are produced in a colony of ants. In the collections I made from this colony, I had observed two quite different sizes of pupae. In my laboratory, I dissected representatives of both types. The smaller pupae were typical of the worker caste, and the larger ones were developing into winged reproductive forms. Given this information, I could assume that the minimum period of time between the establishment of the colony and the discovery of the body inside the toolbox was approximately 12 months.

I also knew that these ants are fairly aggressive predators and would quickly have consumed any eggs and larvae of both the blow flies and the black soldier flies that were in the toolbox. It seemed that both these species must have completed their development to the adult stage before the ants arrived. I reasoned that approximately 30 days would have elapsed before the body became attractive to the black soldier flies for egg laying, followed by a period of approximately 5 months for all the eggs to complete development to the adult stage. This gave a total of approximately 6 months before the arrival of the ants. This 6-month period would also have allowed the body and the other contents of the box to dry enough to attract this species of ant, which does not form colonies in wet habitats. Since approximately another 12 months must have passed before the colony began to produce the winged reproductive forms, the total period of time that had elapsed between the death and the discovery of the toolbox was approximately 18 months. That was my estimate of the postmortem interval.

Given my estimate of the postmortem interval, the serial number of the pager, and the results of the video superimposition of the skull videos on photographs, the victim was finally identified as a local contractor. A man who had worked for him and was known to have been with him shortly before he disappeared was arrested and eventually confessed to the murder. He told the following story: When the contractor criticized him and the quality of his work, he "exploded," kicking the contractor repeatedly until he thought he was dead. Then he stuffed the victim into the nearest receptacle, the toolbox, and dumped it alongside the road where it was found 18 months later. Returning to the crime scene, he cleaned up all traces of the murder, and told lies to account for the victim's whereabouts. He even drove the contractor's truck to the Honolulu International Airport and left it parked there. Eventually, the worker moved to the mainland, where he was arrested and then was returned by authorities to Hawaii to face charges. During the trial, the defense claimed that because of his emotional state, the defendant should be convicted only of manslaughter. The jury rejected this plea and returned a verdict of guilty of second-degree murder.

THE ONLY OTHER case I know of where social insects were involved in establishing a postmortem interval was related to me by Wayne Lord in 1990. In that case, the skeletal remains of a female approximately 15 years old was discovered in January in a mixed oak-maple woodlands area in the Cumberland Mountains of Tennessee. With the skeleton was a loose electric cord, suggesting that she might have been strangled, and a pair of denim jeans. Examination of the skull showed weathering lines, indicating that the occipital foramen, the opening at the base of the skull allowing connection of the brain with the spinal cord, had been in a position that offered insects access to the cranial vault. Inside the cranium, which had been cleaned of tissue by maggots, was the nest of a paper wasp, containing approximately 100 cells, all but 6 or 8 of which appeared at first sight to be empty. Closer examination showed that these cells, although capped, were also empty. Since wasps in that area would not be making nests or rearing larvae during the cooler parts of the year, this indicated that the nest had been active during the previous summer. Given the habitat and the type of nest, the wasps were identified as being a species in the genus *Polistes.* Also inside the cranium was the pupa of a species of small dung fly in the family Sphaeroceridae.

Sphaeroceridae are typically not the first group of flies to arrive at a corpse and usually follow the blow flies in the succession pattern. The presence of the pupa of this type of fly indicated to Lord that a normal pattern of succession had occurred during a warm part of the year. In eastern Tennessee, species of *Polistes* wasps normally seek nest sites beginning in April. Since they require sites that are clean and dry, all tissues from inside the skull had to have been removed before April. Further, Lord thought, the maggot mass that had cleaned the skull had to have been formed the previous autumn, before the temperatures dropped low enough to stop insect activity. Altogether this chain of reasoning gave a minimum postmortem interval of approximately 18 months before discovery of the skeleton.

The skeletal material was submitted to a forensic anthropologist and the jeans to a forensic materials specialist. Both arrived at approximately the same postmortem interval estimate as the entomologist. When the victim was identified, she was found to have been missing for almost 2 years.

8

AIR, FIRE, AND WATER

One of the great pleasures of living in Hawaii is being close to the ocean. Every morning I feel refreshed when I look out at the waves and hear the sound of the surf. But in my profession those same waters can cause difficulties, because the ocean can be a barrier to the insects involved in decomposition. I became involved in a case in point in July of 1993. On the fourteenth of that month a 62-year-old man was discovered nude on his boat about half a mile north of the Heeia Kea Pier on the windward side of the island of Oahu. Near the body were two vodka bottles, one empty and the other half full. Recovering the corpse was difficult because of the fairly advanced state of decomposition and the location of the victim below deck on the boat. Assistance was requested from the military, and the Marine Corps Barracks on Kaneohe Bay sent several Marines to help the medical examiner's investigator. The Marines had been told that the body was partially decomposed and that they would need gloves. So they arrived wearing gloves, white cotton dress gloves, hardly appropriate for the task.

The only insects present on the body were maggots of the blow fly *Chrysomya megacephala* and some egg masses, presumably of the same species. The maggots were on the head and neck, the egg masses in the groin area. The most mature of the maggots were second instars 8 millimeters long. I used temperature data from the Heeia Kea Pier in my calculations of the ADH needed to reach that stage of development. These calculations showed that the blow flies would most probably have laid their eggs during the early to mid-morning hours of July 11. The man had last been seen alive on the afternoon of July 9. The precise time of death was never determined, but it doesn't seem reasonable that this man could have remained active on board his boat near a busy pier for over 2 days without being observed by other boats or coming ashore for food. I concluded that the delayed onset of insect activity was due to the water barrier. Over the years, I have observed that blow flies do not easily cross areas of open ocean without being drawn by a strong attractant. In this case, half a mile of ocean would have kept them away from the corpse until the odors of decomposition became strong enough to be detected from a distance. The prevailing winds in the area come from the north, so these odors would have been blown southward toward the shore and would have attracted the flies after a couple of days of decomposition by bacteria.

Not only can the ocean delay the arrival of adult flies on a corpse, but contact with salt water can affect the hatching and subsequent development of maggots, as I discovered in a case that began when I was called on February 7, 1993, to examine a body found in the Campbell Industrial Park on the southern shore of Oahu. The corpse, nude except for a pair of running shoes, was discovered on the rocks below the high tide mark at approximately 12:45 P.M. I was contacted at approximately 4:00 P.M. by an investigator for the medical examiner and arrived at the scene at 4:45 P.M. By that time, the tide had begun to rise and the body had been moved to a higher area. The death looked like a suicide, but because of the location and condition of the body, I had been called. I examined the body and found only egg masses of what appeared to be a species of blow flies. I collected samples of these eggs and took them to the laboratory to see if I could rear them to the adult stage.

The eggs were placed in an environmental chamber at approximately 6:30 P.M. I first saw maggots hatching from these eggs on February 9 at 6:00 P.M., 48 hours or so after being placed in the chamber. Usually, the eggs of this species, which I later identified, hatch between 12 and 18 hours after being laid. But these eggs took much longer to hatch, 30 to 36 hours longer. Since it took so long for the eggs to hatch, I wasn't sure I would be able to rear the maggots to the adult stage, but once given beef liver to eat, the maggots developed normally. When the adults emerged it was clear the species was *Chrysomya megacephala.*

I wondered what caused the delay in hatching. The only explanation I could think of was the effect of exposure to the salt water, and possibly the cooling effect of the waves washing over the corpse. But any cooling effect would have ceased when the eggs were placed in the rearing chamber. So the exposure to the salt water must have been responsible. Since that case, I have encountered several other cases where eggs have been exposed to salt water for periods varying from a few minutes to several hours. Exposures of more than 30 minutes obviously do prolong the time the insects spend in the egg stage, but the duration of the delay is not easily predictable. In most cases, I have found the delay to be at least 24 hours.

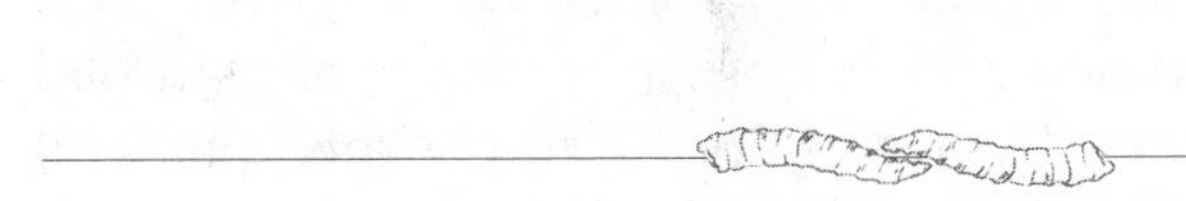

FIRE CAN ALSO affect decomposition rates, but few controlled studies have been done on its effects. I got interested in this topic after attending a symposium sponsored by the FBI in August 1995 on the forensic aspects of arson investigations. When I returned to Hawaii from Virginia, where the symposium had been held, I decided that a research project would be appropriate. At the time I had an undergraduate student in my laboratory, Frank Avila, who was interested in forensic entomology. Together we designed experiments using 50-pound pigs to simulate burning victims that would allow us to detect any differences in the rates of insect invasion of burned and unburned corpses.

For our initial study, we picked a site inside Diamond Head Crater at the Hawaii Army National Guard facility and exposed two carcasses. We burned one of them with gasoline to approximate a level 2 burn on the Crow-Glassman Scale, the scale commonly used to describe burn injuries on human bodies exposed to fire. The scale has five levels, ranging from level 1, with blistering of the epidermis and singeing of hair, to level 5, at which point there is essentially nothing left of the body. In between, in the parlance of my graduate students, there are varying degrees of "crispy critters." At level 2, the body is still recognizable, but shows varying degrees of charring and some loss of parts of the arms and legs.

The other carcass we left unburned as a control. On the basis of anecdotal accounts, we expected a delay in the colonization of the burned carcass. The exact opposite happened. The burned carcass was much more attractive to adult flies for egg laying than the control carcass. Fly activity began shortly after exposure on both carcasses, but the majority of egg laying on the burned carcass was done a day earlier than on the control carcass. As well as laying eggs in the natural body openings, the flies found additional egg-laying spots on the burned carcass where the skin had cracked when exposed to fire. This cracking of the skin had the effect of accelerating decomposition of the burned pig, and insects arrived on the burned carcass 1 day earlier than on the control carcass.

We later repeated this experiment in a rain forest habitat in Lyon Arboretum and obtained even more dramatic results. In this wet habitat the invasion of the burned carcass by flies was even more rapid than in the drier conditions inside Diamond Head Crater. Adult flies even landed on one part of the carcass while other parts were still in flames. This study was recorded by a reporter for the New York Times Video News International. I hate to think what her initial thoughts were on encountering our crew loading dead pigs and gasoline into the back of a pickup truck, but she spent a week visiting the carcasses with us and at the end of it seemed comfortable standing in mud and rain filming us watching maggots feeding on dead pig carcasses. I hope her next assignment was more pleasant.

In the rain forest, the differences in invasion were even more pronounced than at Diamond Head. Most of the egg laying began 4 days earlier on the burned carcass and continued longer than on the control pig. Succession proceeded more rapidly, and the burned carcass was uniformly 4 days ahead of the control throughout the study. As in the Diamond Head Crater study, the burned carcass offered many more egg-laying sites than did the control pig.

These results of course show only what a single level of burning does to a carcass exposed on the surface of the ground. I hope future studies will include other levels of burning as well as investigate the effects on bodies found in different circumstances. Such studies could provide data that would help solve cases in which burned bodies are discovered inside automobiles and buildings, or in other situations where they are not in contact with the ground.

LACK OF CONTACT with the ground is another factor affecting decomposition that is not well documented in controlled experimental studies. I first became aware of such effects when a

Japanese tourist had an unexpected experience on a golf course in Honolulu. Hawaii is host to many tour groups of golfers from Japan, where golf is very popular but the golfers far outnumber the available tee times. These golfers come to Hawaii to play as many courses as possible in a short period of time. When one of these vacationing Japanese golfers teed off one morning, he hit his ball into the rough beside the fairway leading to the sixteenth hole. Hanging from a low branch of a tree in the rough was the body of a man in a late stage of decomposition. The ball struck the skull and dropped to the ground directly under the body. This golfer was exceptionally determined, and once he found his ball, he continued to play the rest of the course. When he got to the clubhouse, he informed the manager through an interpreter that there was a dead body hanging in the rough by the sixteenth hole. The next few parties of golfers were delayed while the scene was processed and the body recovered.

The corpse was unusual. It was largely skeletonized, but the bones were still mostly joined together, partly because the body was fully clothed in a tank top, nylon windbreaker, jacket, and jeans and these garments had helped keep the bones from falling to the ground. Also, it was quite dry in the rough and the body had partially mummified, developing a layer of dried skin in the exposed areas that helped keep the body in one piece.

I was surprised at the lack of any significant beetle activity on the upper portions of the body, since the lower portions of the legs showed considerable depredation by dermestid beetles. As I examined the body and made my collections, I saw that very few different kinds of insects were present. There were empty pupal cases of both the common blow flies, *Chrysomya megacephala* and *Chrysomya rufifacies*, on the outer surface of the upper portions of the body and in the folds of the clothing. Given the normal pattern of development for these two species in Hawaii, the presence of only empty pupal cases indicated a time period of at least 17 days. There were also some larvae of the cheese skipper *Piophila casei* on the upper portions of the body, along with adults of the red-legged ham beetle, *Necrobia rufipes*. The lower legs were infested with adults and larvae of two species of hide beetles, *Dermestes ater* and *Dermestes maculatus.* Of these, *Dermestes maculatus* was the most abundant and had the most mature larvae.

Markedly absent were any insects or other arthropods that would normally crawl, but not fly, onto a corpse. Particularly significant was the complete absence of any of the predators, such as the rove beetles or hister beetles, I was used to finding on a body. In total, I could find only six different kinds of insects on the body. This is an unusually low number for a body that, according to the evidence of the blow fly pupal cases, had been exposed outdoors for at least 17 days. None of the other species on the body yielded any more precise information about time, so my estimate for the postmortem interval was a minimum of 17 days.

When the body was finally identified, my estimate proved to be fairly accurate in spite of the small number of insect species. The man was well known to the police and had last been seen alive 19 days before the discovery of the body. In life he was 5 feet 2 inches tall, but the corpse measured almost 6 feet in length. This discrepancy is what caused the delay in identification; no one that tall was missing at the time.

When the difference in height was discovered, I began to understand some of the departures from normal decomposition patterns I had seen. The only insects that were able to reach the corpse during the first days after death were those that fly onto a corpse. The insects that normally invade the corpse from the soil could not reach it until it had stretched enough to come in contact with the ground. Thus the blow flies were able to begin their activities shortly after death while the dermestid beetles were delayed. Once the feet touched the ground, the dermestids began feeding, but at the opposite end of the body from the head, where like the flies they would normally begin eating. Reversing their usual pattern, they started with the feet and worked their way upward. So much was clear, but I still wondered about the absence of predators, and I couldn't figure out why the blow flies had not removed as much flesh as I had come to expect.

I decided to conduct an experiment to see what would happen during a hanging. Thanks to Wayne Lord of the FBI, I was able to hang a pig under controlled conditions during one of the body recovery courses held at the FBI Academy. The results of the study explain what happened to the corpse on the golf course. I used two pigs, one set out on the ground and the other suspended from a tree. Decomposition proceeded normally in the first pig, and by the end of a month it had been reduced to a skeleton. The hanging pig presented a completely different picture. The blow flies arrived on schedule and laid their eggs around the natural body openings in the head. The flesh flies also arrived on schedule and deposited their larvae in the same places. As the blow fly eggs hatched, they formed a maggot mass and began to feed. But then I began to see a significantly different pattern from that shown by the pig on the ground. When the maggots on the hanging pig circulated to the outside of the feeding mass to cool down and digest their food, they had nothing to hold onto and fell to the ground beneath the carcass. And then they had no way to get back to the food source. So the hanging pig was continually losing maggots and therefore was not being consumed as fast as the pig lying on the ground.

The hanging pig was also more exposed to the effects of wind and air, and its tissues dried out more rapidly than those of the ground pig. Consequently the hanging pig could not be used as food by the maggots as long as the ground pig. Adult flies continued to lay large numbers of eggs on the ground pig for the first week following exposure, but they laid eggs on the hanging pig for only the first 5 days. Further, after the first 4 days, they laid only small numbers of eggs on the hanging pig. This is why the flesh of the body on the golf course had not been completely removed.

I found predators associated with the hanging pig but not where I had looked on the golf corpse. After the maggots fall to the ground, they are permanently prevented from using the body as a food source unless the feet or some other parts still connected to the body are in contact with the ground. If they can't get back to the body, they must find another food source or die. The food most readily available comes from the decomposition of the body above. As it decomposes, fluids and small pieces fall from the body onto the ground in the "drip zone." I have seen this area established by as early as the third day in experimental hangings. In the drip zone maggots can complete their development, and a pattern of succession occurs that is similar to what happens with a body lying on the surface of the ground. I found the predators, parasites, and other missing components of the decomposition patterns in this drip zone. Although I did not sample from this zone under the body on the golf course, I do not think doing so would have yielded much information. The golfer who discovered the body, apparently upset by the sight of the corpse, had obviously needed several swings to hit his ball out from under the body, and the contents of the drip zone were scattered across the course and on his clothing.

9

Drugs and Toxins

Sometimes corpses are not discovered until they have decomposed to the point where all the tissues that would normally be analyzed to help determine the cause of death have disappeared. Especially when it is suspected that drugs or toxins have contributed to the death, the maggots collected from the corpse can be analyzed in just the same way as a sample of human spleen or other tissue. One of the first documented uses of maggots as alternate specimens appeared in a case study published in 1980. The corpse in this case was that of a 22-year-old woman found in a nearly skeletonized condition along a creek bed. The soft tissues had disappeared, and only shreds of skin clung to the posterior part of the body. But there were still large numbers of maggots associated with the body, and these were identified by an entomologist as those of a blow fly, the secondary screw-worm fly, *Cochliomyia macellaria*. With the remains were found a pocketbook containing identification papers and an empty bottle with a prescription label. Dental records confirmed the identification indicated by the papers and the bottle.

The woman had last been seen alive approximately 14 days before the discovery of the corpse. The prescription label on the bottle showed that she had filled it with 100 tablets of phenobarbital 2 days before that. She had a past history of suicide attempts and had left a note strongly suggesting suicide that was found near the corpse.

During the autopsy, maggots had been collected for use in estimating the postmortem interval. Since no soft tissue was available for toxicological analysis, some of the maggots were substituted for the tissues and analyzed for drugs. Sure enough, thin-layer chromatography and gas chromatography showed that the maggots were full of phenobarbital.

Working in France, Pascal Kintz and his colleagues were also testing maggots. In a paper published in 1990, they reported the results of their analyses of maggots and soft tissue samples from several corpses for five prescription drugs. The maggots tested positive for all five drugs, but the analyses of liver and spleen tissues detected only four of the drugs known to be in the corpses.

Also in 1990, Wayne Lord presented a paper at the annual meeting of the American Academy of Forensic Sciences about a case in New England where analyses of maggots for drug content helped determine the cause of death. In this case, skeletonized remains were discovered in February by some people out for an afternoon walk. The corpse was lying face down in a densely wooded area, partially embedded in the frozen soil. There appeared to have been some disturbance of the remains by animals, and the completely skeletonized skull was about 28 feet away from the rest of the corpse. The body was clad in jeans and a plaid shirt. There was an irregular figure-of-eight hole in the pants and right buttock that had obviously been caused by a scavenging animal.

In the morgue, investigators could see that the head and

trunk had been completely reduced to bones. Some flesh still clung to the arms and both legs, and the feet retained decomposed soft tissue and skin. There were many partially decomposed small and large maggots, as well as fly pupae, on the outer surfaces of the corpse, in the body cavities, and within the clothing. There were no tears or cuts to the clothing other than the hole caused by postmortem animal disturbance, and there was no evidence of antemortem injury to the body. The maggots and skeletal muscle from the legs were analyzed, and both tested positive for cocaine.

When the body was identified, as that of a 29-year-old man who had been reported missing in September of the previous year, the presence of cocaine began to make sense. The man was an intravenous drug abuser and had last been seen alive by his girlfriend in her apartment. She believed he had injected cocaine before coming to see her. During the visit, he had installed an air conditioner in her apartment. He had turned on the air conditioner to test it and, when the room got cold, he had become upset and left. He had been seen staggering away from her apartment house, but was not sighted alive again. On the basis of this information and the positive tests for cocaine in both the muscle tissues and the maggots, the medical examiner concluded that the man died of hypothermia while intoxicated with cocaine.

DRUGS ARE NOT the only thing maggots may ingest when feeding on a corpse, as I was to discover in a case that began in February of 1987 when I received three vials of maggots from a medical examiner in Honolulu. The maggots had been collected from the body of an apparent suicide, a 58-year-old man who had a history of suicide attempts. His body, in an advanced state of decomposition, was discovered in a crawl space under his mother's house. He had once tried to commit suicide by shooting himself in the head. This time he had succeeded by using poison, the insecticide Malathion. Beside the corpse was an 8-ounce bottle with approximately 6 ounces missing.

The maggots had been collected from the neck, torso, and head, where they were especially numerous in the nose, mouth, and eyes. The maggots had been preserved in ethyl alcohol, but not fixed before preservation. I was out of town when they arrived, so they remained unfixed in the preservative for a few days until my return. The maggots in one of the vials were too bloated to be used as a reliable gauge of the postmortem interval, but I could identify their species. In all the vials I found maggots of only two species of blow flies, *Chrysomya megacephala* and *Chrysomya rufifacies.* There were only late third instar maggots of *Chrysomya megacephala,* but there were both second and third instar maggots of *Chrysomya rufifacies.* This assortment of stages, given the ambient temperatures in the area where the corpse was found, was what I would anticipate for a postmortem interval of approximately 5 days. But this period was not consistent with the 8 days that had elapsed since the man had last been seen alive and since the date on the receipt for the bottle of Malathion.

I thought it strange that large numbers of maggots were present in the mouth of a body that had ingested an insecticide, especially an organophosphate insecticide such as Malathion. The organophosphates act on the nervous system by interfering with the acetocholine-acetocholine esterase system, which controls the transmission of impulses across the nerve synapses. Exposure to the insecticide causes nerve poisoning. In the case of an adult insect, it first becomes restless and excited, then experiences tremors and convulsions, and eventually dies. Among the organophosphates Malathion is generally regarded as one of the safest because of its low toxicity in mammals. In rats it has a low acute oral lethal dose—the dose that will kill 50 percent of the test animals to which it is administered acutely (in one short exposure)—of 900 to 5,800 milligrams per kilogram, and it can be broken down by the mammalian liver. With its low toxicity to mammals, its ease of application, and its ability to kill a wide variety of insects and mites, Malathion has become a popular insecticide for use in homes and gardens, despite its repulsive odor, which smells to me like cat urine. That the suicide in this case was able to consume 6 ounces of this foul-smelling liquid shows how determined he was to die.

The medical examiner had taken samples of tissues from the corpse at the autopsy and sent them for analysis. Malathion was found in the fat and gastric contents, but not in the blood, the urine, or the fluid from the chest cavity. The intense feeding by the maggots in the mouth area, where there was a lot of Malathion, led me to submit some of the maggots from that area for analysis to the Department of Agricultural Biochemistry at the University of Hawaii at Manoa. The analyses showed Malathion in the maggots at levels that were substantially higher than I expected: 2,050 micrograms per gram. These concentrations were much higher than those needed to kill adults of either species of blow flies, and within the established lethal dosages for rats, yet the maggots appeared to be developing normally. So I looked for data on the toxicity of Malathion or any other organophosphate compounds to maggots. The only information I could find was a study done by Y. Inoue on the effects of Malathion on maggots of the flesh fly *Boettcherisca peregrina.* His research showed that very high concentrations of Malathion were required to kill maggots when the insecticide was applied to the maggots' outer cuticle. But these findings did not explain what happens when maggots ingest Malathion.

Although I was still baffled by the ability of the maggots to tolerate Malathion, I returned to consideration of the insecticide's effect on the adult flies and other insects to resolve the discrepancy between the 5-day postmortem interval seemingly indicated by the insect evidence and the 8-day interval indicated by the other facts in the case. When I consulted decomposition studies, I saw that there were fewer species on this corpse than would normally have been expected. The corpse was outside, in contact with the soil, and, although in a crawl space, open to insect invasions. After a corpse has been exposed for 8 days, I normally find other species of flies, particularly the house fly, *Musca domestica,* and the chicken dung fly, *Fannia pusio,* as well as flies in the families Milichiidae and Sphaeroceridae. In habitats similar to that where this body was discovered, I would also expect to find adult hide beetles in the family Dermestidae beginning to frequent the corpse, along with the predatory rove beetles and the hister beetles. But I had found only the two species of blow flies. I concluded that the Malathion had delayed the invasion of the corpse by the insects for up to 3 days. After that, although some Malathion was still present, the other processes of decomposition had progressed to the point where the body was attractive to the flies in spite of the insecticide. From then on, I thought, the pattern of succession had probably proceeded normally, beginning with the arrival of the two species of blow flies at the natural body openings in the head. But the body had been discovered before other species of insects would have arrived and begun colonizing the corpse.

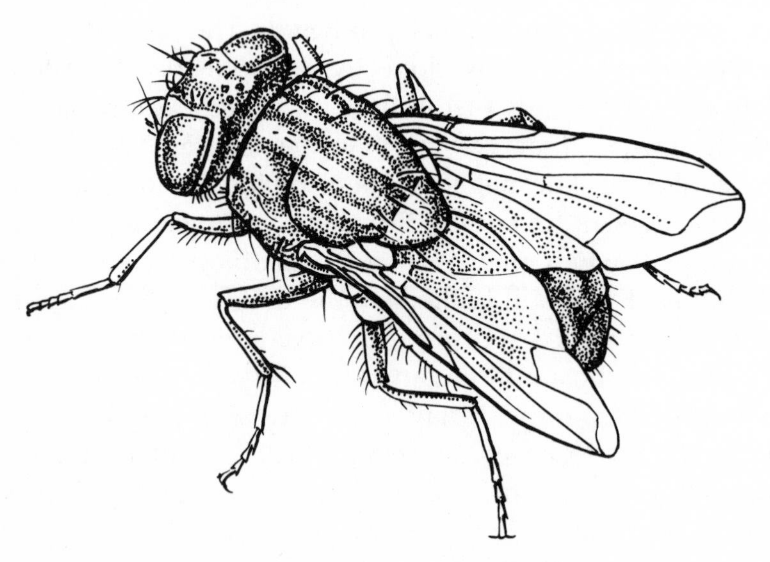

The common house fly, *Musca domestica*

MY EXPERIENCES IN the Malathion case and in other cases I encountered in the 1980s where drugs, especially cocaine, appeared in the maggots associated with corpses made me increasingly curious about the effects of drugs and toxins on the rates and patterns of development of maggots. I had a great deal of information about the rates of development of maggots at different temperatures, in different densities, and under different environmental conditions, but virtually no data on how drugs and toxins affect maggot development. The only study I could find that was at all relevant was the report of some research conducted by Pekka Nuorteva and his colleagues in Finland. They showed that maggots that fed on fish contaminated with mercury had a higher concentration of mercury in their tissues than did the fish and that they had difficulty in moving and entering the pupal stage. Nuorteva and his colleagues then fed the mercury-contaminated maggots to two species of beetles, a rove beetle and a darkling beetle, both of which accumulated the mercury. The rove beetles, *Creophilus maxillosus,* showed no ill effects. But the darkling beetles, *Tenebrio mollitor,* exhibited decreased activity and progressive paralysis of the legs, similar to the effects of Minimata disease in humans.

Finding no other pertinent information in the literature, I decided to conduct my own experiments. In 1987 I approached the Research Committee of the Pathology/Biology Section of the American Academy of Forensic Sciences with a proposal to investigate the effects of cocaine on the development of a species of flesh fly. This was a joint proposal between my laboratory and

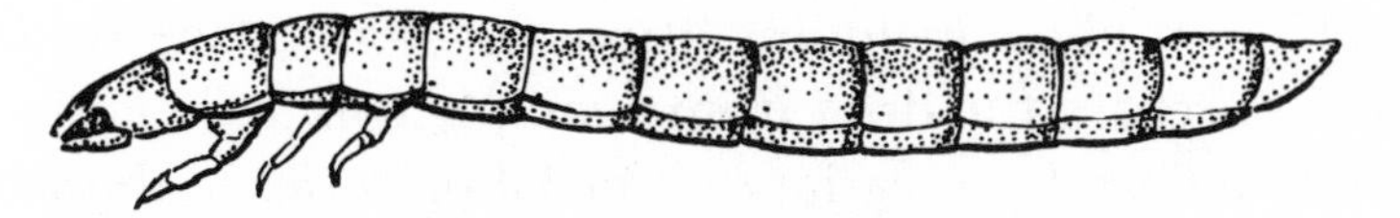

Larva of the darkling beetle *Tenebrio mollitor*

the chief medical examiner for the City and County of Honolulu, Alvin Omori. After some debate, we selected cocaine because it was then the most popular illegal drug and thus the most frequently encountered during death investigations. I selected the flesh fly *Boettcherisca peregrina* as the test fly because I often found it on corpses, it is relatively easily to maintain in a colony, and it deposits live maggots, eliminating the need to wait for eggs to hatch. We were granted the funding and were ready to begin.

To conduct the type of study we envisioned, the drugs had to be administered to a living animal. Cocaine, like many drugs, has no effect until the animal's body metabolizes it into another substance, benzoylecognine in the case of cocaine. If the cocaine is simply homogenized into an artificial culture medium or added to liver tissues, it will not be metabolized and the maggots will not be feeding on the same substances they would encounter in a corpse.

So now we had the money, but we still needed to get the test animals and the cocaine. Once again I found myself in front of the university's Institutional Animal Care and Use Committee. I outlined my proposal and stressed the importance of the work to society. The committee members, remembering my previous requests for pigs, were somewhat taken aback by this new request to give cocaine to rabbits. They asked if I would give the rabbits a tranquilizer to diminish any discomfort or anxiety the cocaine might cause. I explained that I was trying to determine the effects of cocaine on the development of the maggots and that administering an additional drug would confuse the results. Finally they agreed to approve my research plan.

Now I faced the problem of legally obtaining the cocaine. I did not then have a license to work with controlled substances. But fortunately the medical examiner did, and we were able to purchase the cocaine from a chemical supply company. In some of our later studies, including those on heroin, we could not afford to purchase the drug from a supply company and had to rely on donations from various crime laboratories in the state. As we widened our studies to include more drugs, we were required to upgrade the license. Now we probably could legally obtain substances it is illegal for anyone either to produce or to sell to us. Over the years, I have had some very interesting conversations with various agents from the U.S. Drug Enforcement Administration—the last one started with "Oh, you again"—and I have often wondered how I would explain why I was carrying heroin, cocaine, or some other illegal substance if I was pulled over on my motorcycle for a routine traffic violation.

IN ALL OUR studies on drug effects, we followed the same general plan. We used four rabbits. One served as a control and received 10 milliliters of normal saline via an ear vein. The other three rabbits received the drug to be tested in 10 milliliters of normal saline via the ear vein. One of these three rabbits got about half the normal lethal dose, the second a lethal dose, and the third twice the lethal dose. All these doses are calculated by weight. The rabbits receiving the lethal and twice lethal doses die as a result of the actions of the drug; the other two rabbits are sacrificed in a carbon dioxide chamber. After the death of the animals, I remove the livers and take a sample for analysis for drug content. The liver tissues are then exposed to a colony of the flesh flies to establish the four test colonies.

Once the colonies are started, the tedious work begins. At 6-hour intervals, I have to measure 10 maggots selected randomly from each of the four colonies. I continue doing this until all the maggots have completed their development and begun to pupate. Once all have pupated, I check the colonies at 6-hour intervals for emerging adult flies. I have tried scheduling sampling sessions at various times of day, but have never found any schedule that does not leave me and the students assisting me exhausted by the end of the test run. When all the measurements and times have been recorded, I am left with a series of numbers that needs to be analyzed. For the cocaine study, I made up a slide showing all the raw data. It has proven to be quite useful over the years since it has so many rows and columns of numbers that no one can easily make any sense out it. I can talk about virtually any topic, including the national debt, using that slide and no one will have any idea that it's not related to the topic. Without computers, I am sure I would still be working on the first study.

The results of our initial study were quite promising. When I reduced all the figures on the chart to a simple graph, it clearly showed that there was a statistically significant increase in the rate of development of maggots feeding on tissues from the rabbits that had received lethal and twice lethal doses of cocaine. The variations in rates of development were enough to make a difference of up to several days in the estimate of the postmortem interval if the entomologist did not know the drug was present. Once the maggots pupated, the effect seemed to disappear, and there was no change in the time required for development from pupa to adult fly.

It's interesting to compare the effects of cocaine on the maggots with its effects on humans. Humans who take moderate doses of cocaine initially become euphoric and quite talkative. This latter effect is of course not seen in maggots. A human being displays increased energy, appears more alert, and becomes hyperactive. The heart rate increases, up to 110 beats per minute. As the drug is metabolized by the body, these effects diminish. With the maggots, only feeding behavior changes. As the cocaine increases their level of activity, they feed more rapidly, thus ingesting more of the drug, which in turn increases the rate of development throughout the larval stages. But when the maggot reaches the post-feeding third instar stage, feeding and therefore ingestion of the cocaine ceases, the cocaine in the maggot is metabolized, and by the beginning of the pupal stage, the rate of development has returned to normal.

Similar experiments I conducted with heroin, methamphetamines, amitryptyline, and phencyclidine yielded varying results. All these drugs caused some deviations from the normal developmental patterns. Most of them caused some change in the rate of development; with others, such as phencyclidine, larval development proceeded normally, but there was an exceptionally high rate of mortality during the pupal stage. These findings reinforced my belief that to accurately estimate a postmortem interval, I needed to have as much information as possible about any drugs or toxins that might be present in the body tissues.

EARLY ONE MORNING in 1988 as I was completing my studies on cocaine, I received a telephone call from Paul Catts in Spokane, Washington, asking for advice on a case from the Spokane area. The entomological evidence he was analyzing did not appear to make sense and he had run out of ideas.

The case began when the shirtless body of a woman approximately 20 years old was discovered in a wooded area outside of Spokane at about 3:00 P.M. on October 12. The corpse was lying face down in a clear area surrounded by Ponderosa pines and near a dirt logging road. There were several stab wounds on the left side of the chest. The investigators interpreted the blood stains at the scene to mean that the victim had been stabbed while lying on her right side and then rolled over to the face-down position in which the body was discovered. The corpse was in the early Bloated Stage of decomposition, with the face and upper body blackened and the marbled appearance of the arms and legs that is caused by the breakdown of blood vessels.

The body had been refrigerated for 5 days before investigators collected the maggots and submitted them to Catts. He reared these maggots to the adult stage on beef kidney in a container with sand, and from the adults he was able to identify two species of blow flies, *Cynomyopsis cadaverina* and *Phaenicia sericata.* Both these species are commonly found on corpses in the state of Washington. *Cynomyopsis cadaverina* usually lays eggs on a corpse during the first day or two following death. *Phaenicia sericata* typically lays eggs on the corpse during the first day after death and prefers to lay eggs in bright sunlight. The location of the corpse in this case was perfect for that species.

What was puzzling Catts was the stages of development of the maggots. The majority were from 6 to 9 millimeters long and, given the climatic conditions, represented a period of approximately 7 days of development. Most of the other maggots were smaller, of a size consistent with several days of egg-laying activity by the adult flies, activity that was supported by the presence of egg masses in the corpse's hair. Then came the problem maggots, those that were 17 to 18 millimeters long and, given the local temperatures, would have required approximately 3 weeks to reach that size. This time period did not seem plausible given the non-insect evidence in the case.

Catts's first thought was that these very large maggots might have migrated onto the corpse from another nearby corpse, such as that of a animal. But he abandoned this idea after a search of the area failed to find any carrion, and after realizing that any migration from another source would probably have involved more of these giant maggots than had been collected. Stumped, Paul called me to see if I had any ideas. When he phoned I was sitting at my desk looking at a printout of results from my cocaine study. I asked if there was any evidence of drugs in the body. He didn't know but said he would check and get back to me. I suggested that he inquire specifically about cocaine, largely because that was the only drug for which I had any data. In about 45 minutes, Paul was back on the phone, asking how I knew that cocaine was involved and what did that mean. I was tempted to give a long explanation of my superior powers of deductive reasoning, but settled for the truth—I made a lucky guess.

Armed with the temperature data and the information that the victim had been known to snort cocaine, I applied the growth curves obtained from my experimental study to an actual case. The adjustment in the growth curve for a near lethal dosage of cocaine for a human gave a developmental time for the largest maggots of approximately 7 days, the same as for the most numerous maggots present on the corpse. But how had the maggots been exposed to that high a concentration of cocaine? There was cocaine in the body tissues, but not at that concentration. The answer came when we found out how the victim had taken the cocaine, via the nose. All of the maggots 17 to 18 millimeters long had been collected from the mouth and nose. As the facts of the case emerged, it was discovered that the victim had a history of cocaine abuse and had been seen snorting cocaine just before her death. Clearly, the large maggots had fed in the nasal area, where there was a much higher concentration of cocaine than in the rest of the body, and thus had developed more rapidly than the others.

THE PUBLICATION OF papers based on research conducted in my laboratory on the effects of cocaine and heroin interested a number of other entomologists and pathologists in the topic. One of the thrusts of subsequent research has been to find a way to tell how much of a drug was ingested by the human by analyzing the maggots. A major contribution to this work was made by Francesco Introna in Bari, Italy. He fed maggots on liver tissues from 40 corpses in which heroin in the form of morphine had been detected during the autopsies. Introna was able to identify morphine in these maggots through the use of radioimmunoassay techniques. He found a strong correlation between the concentrations of the morphine in the maggots and the liver samples on which they had fed. These results were highly suggestive, but they fell short of being statistically significant. To date, a technique for determining the amount of a drug taken from analyses of maggots has eluded researchers.

My own excursion into attempting to determine concentrations of drugs in a body through analyses of insects came in a case from New England, referred to me by Edward McDonough, a medical examiner from Connecticut. He was faced with the mummified body of a middle-aged woman that had been discovered inside her residence by a real estate foreclosure agent. In examining the area around the corpse, investigators discovered several containers for prescription medications, most of which were empty. The labels listed a number of different drugs, including ampicillin, doxycycline, Ceclor, erythromycin, Lomotil, Elavil, pentazocine, and Tylenol 3. There were also some ovoid blue tablets labeled "Rugby 0230."

The corpse's exterior consisted of mummified skin, some adipocere, a waxlike substance produced by the hydrolysis of tissues, and large quantities of insect feces, which form long threads. Internal examination showed a number of mummified organs, many with extensive insect damage. Inside the stomach there were small maggots, along with an unidentified granular material. There were no wounds on the corpse and no sign of any significant disease that would account for the death. Toxicological analyses showed that there was a lethal concentration of amitriptyline and nortriptyline in the stomach contents and in the desiccated brain. Smaller concentrations of diphenhydramine and cocaine were also present.

McDonough took samples at the scene and during the autopsy of empty fly pupal cases, cast larval skins of hide beetles, and the insect fecal material from the corpse. He sent these samples to the FBI laboratory in Quantico for analyses for amitriptyline and nortriptyline. The fly pupal cases were identified by entomologists at the Smithsonian Institution as belonging to the phorid fly *Megaselia scalaris,* and the cast larval skins from the beetles were of the hide beetle *Dermestes maculatus.*

At the time, I was conducting studies on the effects of amitriptyline, one of the tricyclic antidepressants, in decomposing tissues on the development rate of the flesh fly *Parasarcophaga ruficornis.* I had a supply of empty pupal cases from maggots that had fed on tissues containing known amounts of amitriptyline and nortriptyline. I sent samples of these empty pupal cases and the data from my rearing studies to the FBI laboratory. These samples were subjected to the same extraction and analysis techniques as the samples of pupal cases and larval skins from the woman's body.

One of the major problems in using cast larval skins or empty pupal cases for analyses for drugs is that the drugs or toxins are locked in the cuticle. Maggots, being soft themselves, can be treated like soft human tissues, such as liver or kidney. But to analyze an empty pupal case, the tough chitin/protein matrix must first be broken down to release the drug or toxin. Chemists in the FBI laboratory found that they could accomplish this with either a strong acid or a strong base; and then they could use routine drug screening techniques to isolate the substances. They succeeded in isolating both amitriptyline and its metabolite, nortriptyline, from the cuticle of the pupal cases and larval skins.

The concentrations of amitriptyline in the empty pupal cases both from the woman's body and from my laboratory were higher than the concentrations found in either the cast larval skins of the hide beetles or the insect feces taken from the body. This discrepancy was due to the differences in the food preferences of maggots and beetles. The maggots of both the flesh flies and the phorids feed on the softer tissues, where the concentrations of a drug or toxin ingested as an acute dose are more likely to be found. The hide beetles are not interested in feeding on a corpse until it as been reduced to dried skin and cartilage, and thus do not feed on the tissues that contain significant amounts of drugs ingested in an acute dose.

We had hoped to correlate the concentrations of drugs in the empty pupal cases and the cast larval skins with the concentrations in the different types of body tissues available, but we were not successful. Our results did show that the ratios of amitriptyline to nortriptyline in the pupal cases were the same as those in the dried tissues of the brain and the stomach contents, and these were consistent with what would be expected in a death due to an acute dose of the drug. The same results were obtained for the specimens that I had reared under controlled laboratory conditions with known amounts of amitriptyline in the tissues provided as a food source. This work gave investigators a new tool for use in determining the cause of death in cases involving drugs. Because of the durable nature of the insect cuticle, empty pupal cases—and the drugs and toxins that have been incorporated into their structure—will remain at a crime scene for years after the death and will be relatively unchanged by the elements. Now that a method has been developed for analyzing the substances in pupal cases, investigators will be able to detect their presence years after death.

IN THE LAST few years, the introduction of designer drugs has complicated drug screening and postmortem toxicological examinations by dramatically increasing the number of substances that must be considered. One of the more recent I have dealt with is 3,4-methylenedioxymethamphetamine, which goes by the street names of Ecstasy, XTC, Adam X, and MDM. There is no accepted medical use for this substance and it has a high potential for abuse. It is a relatively new drug to the Hawaiian Islands, but has been widely reported from other areas of the world. In some respects, this drug has presented me with a different picture than have many of the other abused substances I have tested. The maggots grow significantly faster and have a lower mortality rate when fed on tissues from a rabbit given a twice lethal dose of the drug, and also have a lower mortality rate than those in the control colony or maggots fed on tissues from rabbits given lower doses of the drug. The same is true for the mortality rate during the pupal stage, although the time required remains about the same for all the pupae to develop into adults. This is a different pattern from that presented by the methamphetamines I have tested, and shows that I cannot accurately predict the effects of these drugs on estimations of postmortem intervals. There is still a tremendous amount of research to do in the field, and given the current increase in drug use in the United States, it is of major significance.

10

Coping

One of the first questions I get from the audience following one of my presentations on forensic entomology is, "How can you deal with this? Don't you have nightmares?" As I often reply, I learned very early in my career that I had to distance myself emotionally from my cases. If I do not do this, not only can I not deal with gruesome death scenes, but a bias can creep into my work. One of the worst things that can happen in any investigation is for me—or any other forensic scientist—to become an advocate for anyone. I am not a member of any law enforcement agency, and my task is to provide an objective analysis of just one type of evidence. I do the best job I can and leave other areas of the investigation to those who are better qualified to pursue them. At all costs, I must resist the temptation to get involved beyond my field of expertise and adopt a "get the bad guy" mentality.

My experiences in the army have been of some help in dealing with death on a routine basis. During my two years in the army's pathology laboratory, I assisted in many autopsies, but these were of course cases where death occurred elsewhere and I was only seeing death in the calm, sterile atmosphere of the morgue. So my first experiences with crime scenes were unsettling. At the crime scene there is usually evidence of extreme violence and little to isolate me from the outcome of the events surrounding the death of the victim.

I usually manage to cope by dissociating myself from the fact that I am working on what was once another human being. I attempt to maintain scientific detachment and view the corpse as a specimen to be examined rather than as a person who has, in most cases, spent the last minutes or hours of life in pain and terror. The fresher the corpse, and thus the greater the resemblance to a living person, the harder it is for me to remain dispassionate. Fortunately for me, many of the corpses I see are so decomposed that they look more like some of the specimens I worked on during my general zoology laboratory class many years ago, only larger.

I try to maintain my detachment by concentrating on the intense insect activity taking place on and around the corpse. In Hawaii, there are over 200 different kinds of insects and other arthropods that may be associated with a decomposing human corpse. The longer the remains have been exposed to insect activity, the greater the diversity of insects and the more complex the puzzle I have to solve. Each insect is doing something different, and I frequently become so involved in the task of interpreting their activities that I can almost forget they are on a corpse.

But only "almost." I have never been able to completely dissociate myself from the corpses I examine, and indeed if I ever become so accustomed to what I see that I am not disturbed, I believe I would have to find another type of work. I have never met a forensic entomologist who has become that hardened. I do know of colleagues who have become overly involved in a case, to the detriment of their own mental health. We forensic entomologists are not involved in prevention, except to the extent that solving one crime may prevent the same criminal from claiming another victim. I always try to learn from the past, but do not necessarily dwell too long on it.

Of all the coping mechanisms, a warped sense of humor has helped me the most. In spite of the gravity of the tasks at hand, humor always seems to creep into death investigations. This is difficult for an onlooker to understand, but laughter and jokes help release tensions that would otherwise be unbearable.

But even with all these methods of coping with what I see in the course of my work, I cannot distance myself much from some cases, especially those involving children. In the case I discussed earlier of the skeletal corpse of a 30-month-old child found in a shallow grave, the child was still wearing a pink windbreaker and pink running shoes. These were the same sizes, the same color, and the same brand as clothes my youngest daughter owned. I had not anticipated my reactions to these items, but when I went home that evening I decided my daughter should have a new jacket and new shoes.

WHEN THE VICTIM is still alive, I find it almost impossible to dissociate myself. In the vast majority of my cases, I deal with flies that invade a corpse. But not all flies wait until the victim is dead; some species feed directly on living tissue, a phenomenon called myiasis, the invasion of live tissues and organs by maggots.

Feeding on live tissue has been documented for numerous species of flies, most of them blow flies and flesh flies. The need for live flesh varies from species to species. Some species seem to ingest living tissue accidentally, because it is close to something else they feed on, such as dead tissues in a wound or other rotten substances. Others cannot complete their development without feeding on living tissues. Myiasis is of considerable interest to both entomologists and parasitologists. The major book on the subject is *Myiasis in Man and Animals in the Old World,* by Fritz Zumpt. Zumpt proposes two routes for the evolution of myiasis: saprophagous and sanguineous. The species following the saprophagous route began as scavengers and carrion feeders and evolved to the point where myiasis was a necessary part of their life cycle. Those following the sanguineous route were originally predatory species living within the nests of their hosts, and progressed from eating their hosts' blood to eating their tissue. In both cases, Zumpt thought, the original feeding on living tissues was accidental.

Some species, such as the screw-worm flies in the family Calliphoridae, are serious pests of livestock and occasionally humans. In their textbook *Entomology in Human and Animal Health* Robert F. Harwood and Maurice T. James describe a human case dating from 1883. The primary screw-worm fly *Cochliomyia hominivorax* laid her eggs in the nose of a man in Kansas while he was asleep. His first reaction to the infestation was to develop the symptoms of a severe cold. Later, as the maggots fed on the tissues in his nose and head, he complained of an intense irritation in his nose and head, and became slightly delirious. As the infestation progressed, the maggots fed on the tissues of the soft palate and his speech was impaired. Physicians tried to remove all the maggots and did extract over 250. At this point, the victim appeared to be recovering, but ultimately, following an invasion of his Eustachian tubes by the maggots, his condition worsened and he died.

Most invasions of the host are relatively straightforward, but one species has a truly bizarre way of infesting its host. The human bot fly, *Dermatobia hominis,* found in Mexico, Central America, and South America, parasitizes a wide variety of mammals, including humans. The adult flies live in forests, far from any direct contact with the hosts for their maggots. When the adult female is ready to lay her eggs, she captures another fly, usually a blood-feeding species, and attaches her eggs to its thorax or legs with a gluelike secretion. These eggs are attached in such a way that the end of the egg through which the maggot must emerge when it hatches points down toward the next host. When the blood-sucking fly lands on a mammal to take a meal, the maggot emerges and burrows into the host's skin. There it completes its development in an abscess just under the skin. The mature third instar maggot then emerges from the skin to pupate on the forest floor and begin the cycle all over again. When the host is a human the maggots usually do not cause major discomfort.

I first encountered this species in an emergency room in Honolulu. The patient had been vacationing in South America, and shortly after his return, he noticed two lesions on his face that appeared to be infected. In the emergency room, the physician pressed on one of the lesions and a maggot three-eighths of an inch long popped out. By the time I arrived, the patient was fairly calm, but the physician was very upset.

NOT ALL INVASIONS by maggots are detrimental. The beneficial effect of maggots was first recorded by Napoleon's battlefield surgeon, Lavrey in 1799. He observed that soldiers who had been wounded in battle but left on the field long enough for maggots to develop in their wounds had a greater chance of recovery than those who received immediate medical attention. The explanation is simple. The maggots involved feed only on dead tissues, and they remove those tissues from wounds more efficiently than any physician. In addition, the allantoin excreted by the feeding maggots aids in healing and preventing infections.

The first deliberate introductions of maggots into wounds for cleaning are attributed to one Zacharias, a military physician for the Confederacy during the Civil War. The use of maggots as a form of therapy continued for some period of time thereafter. I have a reprint of a paper on maggots used to remove dead pulp from root canals that was published in *Dental Survey* by George C. Dreher in 1933. Eventually maggot therapy fell into disfavor for obvious reasons, but maggots have recently begun to be used again in hospitals in the United States to clean badly infected wounds, an effort championed by Ronald A. Sherman of the Veteran's Affairs Medical Center in Long Beach, California.

Aside from the medical assistance maggots may provide in cases with neglected wounds, there is a forensic application to their activities in living victims. Wayne Lord described a case in which several children were brought into a hospital emergency room suffering from severe diaper rash, malnutrition, and general neglect. During the physical examination, maggots were discovered in their anal and genital areas. The maggots were collected and preserved for examination by a forensic entomologist. His findings showed that the maggots had been developing for 4 to 5 days; this was then presumed to be the period of time that had elapsed since the children had last had their diapers changed. In this case, the entomological information was the only evidence of the length of time the children had been neglected, and the entomologist's findings weighed heavily during the hearings to determine the children's future.

Like children, the elderly and infirm are at risk of being neglected and abused. I have been involved in several cases where people in care facilities have been discovered suffering from severe bed sores that contained well-developed maggots. The maggots in these instances are beneficial to the patient because their activities reduce the chance of infection and gangrene. Forensically, the maggots document the duration of the neglect. If a bedridden person is bathed every day, as the caretakers in these cases claimed, it is extremely difficult to explain the presence of 5-day-old maggots in the sores.

Neglect of the elderly who are staying with family members is even more disturbing and hard to stomach because the family is traditionally thought to be the strongest source of support for the elderly. In a particularly disheartening case, I was called to examine the corpse of a woman who had been discovered dead by her family. She had suffered a stroke and was living with two of her grown children and their children. According to her daughter, she was alive when the daughter checked at approximately 1:00 P.M. but was dead when she next saw her at approximately 5:30 P.M. The body was clad in a dress and a diaper and seated in a wheelchair when the investigators from the medical examiner's office arrived. They could immediately see that the body was encrusted with dirt and feces. Later, when the diaper was removed at the morgue, it was found to be filled with maggots, and there was an area of dead, rotting tissue on the lower back, extending even into the abdominal cavity, that contained numerous maggots. I was called in to take samples and provide an estimate of the time required for the maggots to have reached their stage of development.

All the maggots I collected were third instars 9 to 10 millimeters long. I reared a sample to the adult stage and identified them as the blow fly *Phaenicia sericata.* None of the maggots I collected or observed on the corpse had reached the post-feeding stage, and the most mature appeared to be about midway through their third instar development. Using developmental times from published studies and adjusting for normal human body temperature, I calculated that it would have taken a minimum of 32 hours for these maggots to have reached that stage of development.

Published studies of fly life cycles usually give several different measures of the time needed to reach each stage of development. The minimum and maximum periods are the times required for the first and last maggots in a colony to reach a given stage of development. The mode is the time at which the greatest number of maggots reaches a given stage of development, and the mean is the average time required for all maggots in the colony to reach the given stage of development. Virtually all studies give the minimum and maximum times, and most studies also include either the mean or the mode values. In the case of my third instar maggots, the minimum time was not helpful. If I was looking at a minimum time, some of my specimens should still have been in the second instar. But since I had no second instars, the value given for the mode was appropriate for my specimens. This value indicated a period of development of approximately 50 hours, a time consistent with other evidence discovered in the case. Unfortunately, as a result of several legal complications, nobody was tried for the obvious mistreatment of the old woman because the victim died before the neglect was discovered.

I FIND CASES of child abuse especially distressing. One of the worst I have encountered began in April of 1990, when a man out for a walk along the side of Lake Wilson on the island of Oahu thought he heard a small dog or kitten caught in the heavy undergrowth. He sought help from others to search the area. Going to investigate, they found not a dog or a kitten but a 16-month-old girl suffering from dehydration, bruising, and copious mosquito bites. Initially, investigators thought that she had been exposed for about 2 days and would probably have died within the next 24 hours had she not been discovered.

When found, she was wearing pink pants and a diaper. There was an apparent infestation by maggots in both her anal and her genital areas. I was called by the Honolulu Police Department to assist in their investigation. I took the samples of the maggots that had been collected from the child by the emergency room personnel to my laboratory for identification. I also examined the pants and diaper. All along the front of the pants, I found large egg masses of blow flies. In the diaper, I found first and second instar blow fly maggots. I took these maggots to the laboratory to be reared to the adult stage. The first instar maggots were

3 to 4 millimeters long, and the second instars were still in the early part of that stage and were only 5 millimeters long. The eggs collected from the pants were almost ready to hatch, and well-developed first instar maggots were clearly visible inside.

When the adult flies emerged, I identified them as one of our common species of blow flies, *Chrysomya megacephala.* Although I had dealt with this species in almost every case I had worked on up to that time, I had never encountered it in a case of myiasis in Hawaii. Yet these maggots clearly were feeding on living tissues in the lesions caused by diaper rash and in the genitals and anus. Studying publications concerning this species, I found that in other parts of its geographic range it had been implicated in myiasis but, strangely, not in Hawaii. The reported cases were classified as instances of facultative traumatic myiasis, a category that includes cases where maggots invade wounds on animals, including humans, but feed on living rather than dead tissues. I speculated that the female flies had initially been attracted to the feces that had accumulated in the diaper. The eggs laid on the outside of the pants had hatched, and the maggots had moved into the diaper to feed on the feces. Once this food source was consumed, they invaded the sores associated with the severe diaper rash and then moved into the genitals and rectum. On the basis of my laboratory results, I estimated that all this took place over a period of approximately 23 hours.

In addition to the time required for the maggots to develop to the stage they were in when collected from the child, I could determine some other things. I knew this species of fly does not lay eggs on a moving object; even the slightest motion deters the female from laying eggs. Moreover, *Chrysomya megacephala* and related species prefer to lay eggs in the dark—frequently that means the underside of the victim. Given the number of egg masses and their location on the child's pants, I concluded that she had been lying motionless, face down, for at least 23 hours before her discovery.

I was asked to testify in this case before both the grand jury and the subsequent trial jury. As the story unfolded, it became clear that the mother had put the child, named Heather, in a stroller and taken her for a walk along with her older daughter. At the edge of Lake Wilson, she had her 6-year-old daughter take her sister out of the stroller and put her down by the edge of the lake. She then took the older child and left. When questioned by her ex-husband, Heather's father, about his daughter's whereabouts, she told him that Heather had been taken from her by personnel from social services. But since she had already told the police that Heather had been kidnapped by two black men, the jury did not believe her and found her guilty of attempted murder.

This case was appealed up to the Supreme Court of the State of Hawaii. The appeal was based partly on the defense's contention that my testimony had been unduly prejudicial to the jury. The defense said my description of the activities of the maggots was so graphic that the jury was unable to remain objective while they considered the evidence. I imagine that it was indeed difficult to feel sympathy for a mother who had abandoned her child, particularly if maggots were involved. In the Supreme Court, the majority opinion upheld the guilty verdict, although there was a dissenting opinion that the slides I used to explain the entomological evidence and the vials of maggots introduced as evidence were so disturbing that they had prevented the mother from receiving a fair trial. I was much more disturbed by what had happened to the child. She did, I am happy to report, recover, and was later adopted by her aunt.

This case made me realize that an entomologist who did not know about maggot activity before death could easily misinterpret the insect evidence and thus give an inaccurate estimate of the postmortem interval. Before this case, I had assumed that myiasis did not occur with *Chrysomya megacephala* in Hawaii. Now it was clear that under some circumstances this blow fly does begin feeding on the body *before* death. Had Heather died before being found and the maggots been collected by someone else, I could have been off in my estimate of the postmortem interval by up to 39 hours. Without knowing where on the body the maggots had been collected, I would have assumed that the body had been *cooling* during the entire period of the maggots' development, rather than being at a temperature close to 98.6°F, normal for humans. Accordingly, I would have assumed that the maggots had taken longer to develop on the cool body and that the postmortem interval was longer that it would actually have been. But if I had myself examined the corpse, I would have been alerted to the possibility of myiasis by the concentrations of maggots in the genital and anal areas and their absence from the openings of the head.

THE WAY EVIDENCE is collected can present a number of problems the forensic entomologist must cope with. I frequently receive shipments of insects that have been collected from corpses by people who fail to document where on the body, and sometimes even when, they gathered them. I always try to obtain this information, but am not always successful, and then there is a real danger of misinterpretation. It is not unusual for someone untrained in entomology to become overwhelmed by the sheer numbers of maggots on a corpse. In such instances, especially since most maggots look alike, the investigators often do not see the need to separate collections from different parts of the corpse, and all the specimens are sent to me in a single container.

Sometimes an entomologist is even asked to review a case after the entomological evidence has been destroyed. Usually these requests are for an estimation of the postmortem interval based on examination of photographs taken at the death scene or during the autopsy. Most often these requests come from defense attorneys who are desperately trying to find grounds for appeal. Except in very unusual circumstances, I do not make such estimates and advise others against making them. Any estimate of the postmortem interval based only on photographs, especially if the photographs were taken primarily to show things other than insects, is very tenuous. Unfortunately, not all of my fellow forensic entomologists agree with this stance, and I have been involved in several cases where a very specific postmortem interval estimate was given based solely on death scene photographs. My role in these cases has primarily been to demonstrate the limits of such estimates.

What can happen when evidence is not carefully collected at the appropriate times and too much reliance is placed on crime scene photographs is well demonstrated by an investigation that began on September 14, 1988, when the body of a 7-year-old girl was found in a suburban area of the Midwest. The corpse was wrapped in a bedspread, with head and shoulders exposed, and was clad in a nightgown. A rope was wrapped around the girl's neck. She had last been seen alive on September 10, and her body was discovered 4 days later, during the afternoon of September 14.

Although the body was discovered in September, an entomologist was not consulted until December of that same year. The entomologist was allowed to collect specimens from the bedspread and nightgown, which had been stored in a police evidence locker. No insects had been collected at the scene or during the autopsy. The specimens were in very bad condition, shriveled and dehydrated, but the entomologist could identify them to the species level. After examining these materials at the police station, the entomologist was taken to inspect the crime scene. There he scrutinized the area and took soil and litter samples for later comparison with the materials from the case. In addition, he was given 39 photographs taken at the crime scene and during the autopsy, some temperature data, and copies of the medical examiner's report, the toxicology report, and the investigator's report.

According to the entomologist's report, the vast majority of the insects on the materials he examined were maggots, but there were also some predatory rove beetles, *Creophilus maxillo-*

sus. The maggots he collected from the clothing were two species of blow flies, *Phormia regina* and *Phaenicia sericata*, with *Phormia regina* the most numerous. Using these specimens and the scene and autopsy photographs, he determined to his satisfaction that the most mature specimens were third instar maggots in the early portion of the post-feeding stage, just beginning to prepare for pupation. He then presented a series of times representing the minimums required to complete the different stages of development for both species of blow flies under the environmental conditions prevailing at the scene. He did not cite the sources for his data, except to say that they were "averages based on several replicate experiments." After examining the photographs and the specimens he had collected, he concluded that the most mature specimens were about 36 hours into the post-feeding third instar stage. Given these assumptions, it would have required 112 hours for *Phormia regina* to reach that stage of development and 115 hours for *Phaenicia sericata*. His report gave a range of 112 to 115 hours as the estimate of the postmortem interval and 9:00 P.M. on September 14 as the point at which the biological clock had been stopped. Counting back from that point, he declared that death had occurred between 1:00 and 4:00 A.M. on September 10. On the basis of this estimate and other facts in the case, investigators charged the girl's parents with her murder.

I was asked to review the report by the defense attorney because even though he was unfamiliar with entomological evidence and techniques, he thought the estimate seemed unusually precise under the circumstances. I was provided with all the written materials given to the prosecution's entomologist as well as his report. I was not given access to the specimens he had collected from the bedspread and nightgown. As I read through the materials and examined the photographs, I couldn't help thinking that the time limits the other entomologist gave were too narrow to be realistic. In fact, these limits were narrower than I would have expected if the entomologist had made his collections from the body at the scene and during the autopsy. I did not see how the entomologist could have arrived at such precise conclusions with reasonable scientific certainty.

Most of the photographs had been taken to record aspects of the body and surroundings other than the maggots and did not provide enough details to permit accurate identification of either the species or the stages of development. Many of the photographs did not have any scale on them; others did have scales but were so poorly exposed that the markings on the scales were indecipherable. It seemed to me that it would be impossible on the basis of these photographs to state with any degree of certainty that the most mature specimens on the body had been located and measured; yet these pictures were the basis for the estimated postmortem interval. Although I had not seen the specimens collected from the bedspread and nightgown, past experience told me that they would be of very limited use in determining developmental time. They would, however, certainly be adequate for identifying the species, and I had every confidence in the other entomologist's ability to make those identifications. But I had serious doubts that the condition of the specimens justified saying that they were in the early post-feeding third instar stage. If they were in the post-feeding stage, they should have been in the process of leaving the body to look for a safe place to pupate. But the maggots in the pictures were still part of an active, feeding maggot mass, indicating that the specimens were most probably still in the feeding portion of the third instar stage.

I also had doubts about the developmental times the prosecution expert was presenting. The times differed slightly from those in the data sets I used, and indicated a more rapid rate of development than I would have expected. Moreover, these times were being used as minimum times in the analyses, but were also stated to have been "averages" of the times derived from several experiments. I did not understand how a figure could be both a minimum time and an average time for the same experiment. Also, I wished I knew the maximum and minimum developmental times for these experiments. Frequently these values are of more use to me in delimiting a time period than are average values. For example, 9 can be the average of a series of numbers ranging from 8 to 10, but it can also be the average of a series of numbers ranging from 1 to 20. When entomologists estimate the postmortem interval, giving an opinion that has very serious consequences, they should give the range of values.

I was also suspicious of the temperature data used to generate the estimated period of insect activity. These data were from an airport 40 miles away from the crime scene, and there had been no attempt to find out whether there were any differences between the temperatures at the crime scene and those at the airport. Even in Hawaii, where temperatures are relatively uniform, to assume that temperatures taken 40 miles away from a scene are an accurate representation of conditions at the scene requires a leap of faith that I am not willing to make.

Another assumption that appeared to have been made was that all maggot feeding and growth somehow stopped immediately and permanently at 9:00 P.M. on September 14, 1988, when the body was refrigerated. This was almost certainly not the case. Although maggot activity slows and almost ceases for practical purposes shortly after a body is put into a refrigerated crypt, it does not stop immediately, particularly if a maggot feeding mass has been formed. Some development would have continued. When the body is removed from the crypt and begins to warm, the maggots also warm and once again begin their feeding activities. The specimens collected from the bedspread and nightgown in this case were removed and stored separately from the body in an unrefrigerated area. Under such circumstances, if there is still available food, such as tissue fragments or decompositional fluids, the maggots will continue their development until they exhaust the food supply. This would have been the situation for the specimens collected by the entomologist in December. These specimens did not necessarily indicate the stage of development at the time the body had been discovered. I thought that the maggots had continued their development on the bedspread and nightgown while they were stored in the evidence locker.

On the basis of the materials I reviewed, I concluded that the estimates given by the prosecution's entomologist were *possible* times, but by no means *probable* times, and certainly did not give the entire possible range. If I accepted his conclusion that the maggots were 36 hours into the post-feeding third instar stage, my lower estimate would have been 96 hours, rather than his 112 hours. The upper limit of my estimate would have been 140 hours. If I was correct that the maggots were not actually in the post-feeding stage, then the lower limit for the interval would have been 61 hours and the upper 104 hours. Based on my previous experience, these times were consistent with the appearance of the body in the photographs. My estimates did not limit the available suspects to the girl's parents as clearly as did those of the prosecution's entomologist.

I was also still bothered by the unusual precision of his estimated times, which he stated were based on measurements of maggots from the photographs. I had not seen any specimens that I thought justified that level of certainty.

I finished my report and sent it off to the attorney the day before I left for a forensic entomology symposium, where I was to present a paper on patterns of human decomposition and another on the uses of insects as toxicological specimens. The prosecution's entomologist was a scheduled presenter at the same symposium, and his announced topic was the use of blow flies in estimating postmortem intervals. Since we were working on opposite sides of the same case, I had decided to limit my contacts with him during the meetings, and I assumed that he would also be careful to avoid any situations where we might appear to be discussing the case. I had also arranged to meet with the defense attorney following the symposium because he had another case in the city where the symposium was being held. I trusted we would find a way to recognize each other because I had never met him, but had only spoken with him on the telephone.

As I sat in the audience following my presentation, I noticed a man—much better dressed than most entomologists—enter and take a seat toward the back of the room. I paid no further attention to him because the prosecution entomologist was beginning his discussion of blow flies. To my amazement, a couple of minutes into his presentation he launched into a detailed discussion of the case we were both working on. As he showed slides he had prepared from the photographs and explained his analyses, it became obvious that he had no idea I was also working on the case. In his discussion, he answered the question that had most bothered me from the start: Where were the maggots he had measured so carefully for his analyses? It turned out that most of his analyses were based on a *single maggot* in a photograph of the lower jaw.

As he concluded his presentation, hands began to go up in the audience for questions. The first person called upon was the well-dressed man who had caught my attention as he entered. As soon as he spoke, I recognized the voice of the attorney for the defense. He asked several questions and then, as the audience was dispersing, approached the podium. Without apparently giving any thought to whom he might be talking, the prosecution entomologist gave the attorney further details of his analyses. I was disturbed by what was going on and attempted to interrupt the conversation. My efforts were in vain; the prosecution entomologist was talking to someone interested in his work and was not about to be interrupted. After that conversation, the attorney didn't seem to care if he spent much time reviewing my report that afternoon. He already had more than he had hoped for from the trip. All the way back to the interview room, the attorney was whistling and cheerfully saying under his breath, "I've got him. He's mine."

Ultimately, the defense lawyer never had a chance to get him. Once the prosecution lawyers received my report during discovery, they dropped their entomologist's testimony from their case, and I was never called to testify. In the end, the mother was found not guilty, but the father was convicted. Shortly after the mother's release, I am told, she took the returned bail money and departed, leaving her husband in jail and most of the attorney fees unpaid.

11

TESTIFYING

*I*n most cases, entomological evidence can elucidate only part of the puzzle, usually the time since death. But sometimes evidence gathered by entomologists can help to provide an explanation of the circumstances of the death, as in this case from the Grand Canyon.

On August 26, 1992, the bodies of a man and a woman were discovered by hikers at the base of a cliff about half to three quarters of a mile from Tonto Trail on the north rim of the canyon. The bodies were lying about 100 feet apart and the woman had a crude splint on her leg, indicating some mishap had occurred before death. The bodies were removed by a helicopter at approximately 6:30 P.M. and taken to the Coconino County Medical Examiner's Office in Flagstaff, Arizona, where they were placed in a refrigerated crypt with the temperature held at 32°F. The medical examiner conducted autopsies on the afternoon of August 27, 1992. The results of the autopsies indicated that both had died by drowning, even though the bodies had been recovered in a dry, desert area. The two had last been seen alive at approximately 11:30 A.M. on August 18, 1992. Since the deaths had occurred in a national park, federal as well as local investigators quickly became involved.

I was contacted by the regional forensic consultant for the U.S. Air Force Office of Special Investigations and asked to provide assistance in determining the time of death. Fortunately, the Special Agent in charge of this case not only was aware of the potential of forensic entomology but had also received some training in the collection and shipment of insect specimens. I gave some further guidelines and instructions over the telephone and was shipped 12 different samples of insects and other arthropods from various parts of the victims' bodies. At my suggestion, on the day following our conversation, August 28, the Special Agent returned to the scene of death and collected soil samples from the areas where the bodies had lain. These were all shipped to me on August 31.

The specimens arrived in my laboratory via Express Mail on September 1, having spent very little time in the campus mail system or the college or departmental offices. Although all the samples were properly packed, the package still exuded a definite smell, and it was not allowed to remain in any of the offices very long. I put the soil samples into Berlese funnels for 48 hours to allow all of the arthropods time to migrate through the soil and into the preservative solution. Then I turned my attention to the insect specimens.

I found four species of flies in the preserved and living specimens sent me by the agent. Three were blow fly species typically recovered from decomposing human and animal remains. I easily recognized the larvae of the secondary screw-worm fly, *Cochliomyia macellaria,* by their dark brown to black tracheas, and my identification was confirmed when the adults were reared from the maggots. Also among the larvae were specimens of *Chrysomya rufifacies,* a blow fly species common in Hawaii, but a recent invader of the continental United States. The third species, *Phormia regina,* I identified from the reared adults. The last species was a flesh fly in the family Sarcophagidae. I was unable to make a species-level identification of this fly because none of the maggots completed their development to the adult stage. Unfortunately, this is common with the sarcophagids, particularly if they have been exposed to cold temperatures before being allowed to pupate. Overnight temperatures at 32°F are cold for a sarcophagid maggot. Along with the maggots, there were also adult specimens of the large rove beetle *Creophilus maxillosus.* This species preys on the maggots feeding on a dead body and typically arrives early in the decomposition process. It has no interest in the body as a food source, but eats only maggots and any other insects feeding on the body.

I examined the soil samples under a dissecting microscope to recover specimens that might have died before the processing or that had not come down through the funnel. I identified additional maggots of all three species of blow flies as well as specimens of the rove beetle *Creophilus maxillosus.* I also found a smaller species of rove beetle in the sample taken from where the woman's body had lain and several kinds of mites commonly associated with the early stages of decomposition. Hand sorting through the samples, I collected pupae of the blow flies and reared them to the adult stage. These were the secondary screwworm fly, *Cochliomyia macellaria.*

Along with the insect specimens, I had received weather data from several locations in the general vicinity of the site. This information was from the weather stations maintained at the north and south rims of the canyon, a weather station located on the canyon floor, and Phantom Ranch Ranger Station. I wished I could have had a hygrothermograph placed at the site of discovery for several days to determine any differences in temperature between the site and the weather stations. But the site was remote, the approach hazardous, and the helicopter pilot was unwilling to go close enough to place the instrument. The data from Phantom Ranch Ranger Station proved to be the most useful because this station is at an elevation and exposure similar to those of the site. Using the weather data I had and adjusting for temperature variations, I estimated that death had occurred approximately 120 hours before the bodies were put into the refrigerated crypt at the Coconino County Medical Examiner's Office. This is the amount of time required for the most mature maggots, in this case third instar *Cochliomyia macellaria,* to reach the stage of development of the specimens collected from the bodies. According to this estimate, the couple had died some time during the evening of August 21.

But how had the two people died? Drowning is not a common cause of death in a desert environment. Had the couple drowned elsewhere and had their bodies later been transported to this remote location? That seemed unlikely. Why were the bodies discovered 100 feet apart? Had they been moved following death? The answer to these questions lay in the weather data. On the evening of August 21, there had been a heavy rainfall in one area of the Grand Canyon some distance from the site, enough to cause flash flooding in other areas of the canyon.

It appeared that the two hikers, last seen alive on August 18, had left the marked trail through the canyon. The woman had fallen, injuring her leg. Her companion then climbed down to assist her and fashioned the crude splint. Because of her injuries or oncoming darkness, they found themselves spending the night of the twenty-first on the canyon floor. During the evening, a flash flood swept through the canyon, drowning them both, and depositing their bodies downstream, 100 feet apart on rocks at the base of a cliff. As is typical in flash floods, the water evaporated quickly, leaving no trace. With daytime temperatures ranging from 81 to 100°F, their clothing quickly dried and insects immediately began to arrive at the bodies. Thus they were discovered 5 days later, having drowned in a desert.

IF THIS CASE had happened in 1983, when I became involved in forensic entomology, instead of in 1992, my conclusions about the circumstances of the deaths would have been largely discounted. In fact, there would probably not have been any entomological investigation. Insects were not then regarded as a significant source of information by most medical examiners, let alone crime scene investigators or lawyers, except under extremely unusual circumstances. One of the major problems facing forensic entomologists at that time was educating the forensic and legal communities about the potential of insect evidence. Through workshops, seminars, lectures, and memberships in various societies, other forensic entomologists and I have managed to convince these groups that there is a place for an entomologist behind the yellow tape. In retrospect, we may have been too successful. In 1983, entomological evidence was viewed with skepticism. By the early 1990s, the situation had changed and people involved in death investigations were willing to believe almost anything we told them. Entomologists who had never consulted on a single homicide case were conducting training sessions for law enforcement agencies as well as other entomologists. With a number of people having a minimal exposure to forensic entomology entering the field, it was only a matter of time before major problems cropped up.

The goal of every forensic entomologist is to produce a set of carefully analyzed data that can be used in a court of law. A courtroom is about as foreign and hostile an environment for a scientist as can be imagined. I have now appeared in court many times as an expert witness, both for the defense and for the prosecution, and every time I enter the courtroom, I still feel much as if I am leaving the planet. The fact that a homicide or some other crime has occurred seems almost irrelevant to what takes place in the courtroom, crowded with the judge, the jury, the bailiffs, the sheriffs, the attorneys, and the court reporter, who frequently has to ask me how to spell the Latin names of insects. Of course, the suspect is there, usually under some form of guard, but not always. During the course of one case in Tennessee, I had to make several trips between Hawaii and Nashville over a 1-week period. On my last trip I caught the late night flight out of Honolulu and arrived in Nashville at 8:30 A.M. The plan was for me to be taken directly from the airport to the courtroom, to speak briefly to the attorney, and then to get a couple of hours of rest before testifying. When I arrived at the court that morning, the attorney looked a little disconcerted. I asked when I would have to be back in court to testify. He asked if I could go and change clothes in the restroom because there had been a change in plan and I was needed in 15 minutes. I went to change and found myself in the company of another man also changing clothes. We exchanged a few pleasantries before he left. The next time I saw him he was being escorted into the courtroom by the bailiff.

One characteristic of most attorneys I have encountered that I still find remarkable is the ability to mentally move in and out of a case. I am often considered the opposition while I am giving testimony or during cross examination, but the situation changes as soon as a recess is called or the day is over. In the absence of the judge and jury, conversation turns to the game the night before, sports in general, and occasionally politics, movies, and other current events—people even tell lawyer jokes. These conversations often include both the defense and the prosecutions attorneys, the suspect, and the bailiff. But when the jury returns, each reassumes correct courtroom demeanor and the trial continues. This has happened so often that I am no longer surprised when it does, but I'm still amazed.

Academics and the legal system do not usually coexist in comfort. The laws of science and the rules of evidence have little in common. In theory, Academia functions on the principle of collegiality. In theory and reality, the American legal system is adversarial. The average academic entering the legal system is in for a tremendous culture shock.

Until recently the courts had little outside help in determining the competence of a forensic entomologist. For other forensic disciplines—pathology, anthropology, odontology, psychiatry, and toxicology—there were established boards or other regulatory bodies providing certification of at least some level of competence. But until 1996, there was no such board or body for forensic entomology. It was up to each entomologist to provide documentation of his or her competence. This might at first glance seem to be an ideal situation, each person standing on his or her own merits. But in reality, it was a nightmare. Lawyers and judges had no basis on which to evaluate the entomologist's qualifications.

THERE SEEMS TO be a predictable pattern to most ~~court~~ of my court appearances. I am initially contacted either by the police or the medical examiner to examine a corpse and collect specimens for analysis. I perform the analyses and give a written report to the medical examiner or the police as soon as possible since there is always great urgency at that point in the investigation. With only rare exceptions, I then drop out of the investigation. Usually I file the case and tend to forget most of the details.

Then at some undeterminable point in the future, I receive a visit from either a police officer or a sheriff's deputy with a subpoena. These visits almost always occur during the evening when my wife and I are entertaining at home or during one of my lectures at the university. One quick look at the subpoena usually tells me absolutely nothing about the case. The subpoenas give the name of the accused, the police case number, and the name of the attorney in the prosecutor's office I am to contact. Since I use only my own case numbers, the medical examiner's numbers, and, occasionally, the name of the victim, subpoenas don't tell me much of anything except the date I am supposed to appear in court. Experience has shown that this is the one day I will not be appearing in court to testify. Once I contact the prosecutor, I discover what case I am actually dealing with and usually find that the case has been postponed for a period ranging from a couple of weeks to a year.

When a trial date is set, I meet with the prosecutor or the defense attorney to review the evidence and explain the principles behind my analyses. We go over the case in detail and cover those parts of it he or she feels are particularly significant. I explain my reasoning to the attorney and the limits of my data and estimates. There is of course another side to the case, and the process of discovery, or the exchange of information between sides, is required. If asked to, I will then meet with the attorney from the opposing side. These are always very interesting meetings. On the one hand, that attorney is going to attempt to show in court that my conclusions are less than credible. On the other hand, the attorney needs to appear professional and not overtly hostile, possibly because such attorneys may need me at some point in the future. I provide the same explanation I have given for the opposing attorney and a mild form of cross examination begins. One of the most mutually productive forms for this exercise to take is for the attorney simply to ask me directly under what circumstances I could be wrong. Recently, another approach is to tell me that another entomologist has reached different conclusions and then to ask me why we do not agree. In several cases, this process has led to agreement over the entomological evidence before we get into court. But often the attorney attempts to "play entomologist," and the results are usually less than successful.

When the case finally comes to trial, I am given a date and time at which I am actually to appear. Since there is a witness-exclusion rule in effect in Hawaii and in most of the jurisdictions in which I have testified, I wait in either the hallway or a witness waiting room until it is my turn to testify. Once called, I am administered the standard oath, which is only slightly less encompassing than a marriage vow and is designed to assure that I will to the best of my knowledge and ability present a truthful account of my interpretations of the evidence I have analyzed. But this account is governed by the rules of evidence, which at times seem to be deliberately designed to prevent you from telling the entire truth. Questions by the lawyers can be constructed to elicit the responses they desire, even if those answers do not accurately reflect the facts. I am frequently required to answer a very complex question with either a "yes" or a "no." One of my favorite questions is, "Are you absolutely sure that there is no other possible explanation for this?" In biology, as in most of life in general, there are almost always alternate explanations, some valid and some not (the dog ate my homework). Usually, but not always, this type of question is posed by the attorney for the opposition. If the attorney who called me is not paying close attention, the final result of my testimony may be quite different from what he or she hoped for.

Before I testify, my qualifications as an expert witness in the field of forensic entomology must be established for the court. This is now a simpler exercise than it used to be. After the first time I was qualified as an expert in the state of Hawaii, my documented credentials were more quickly accepted and the questions took less time during later trials. During the early 1980s, if insect evidence was used by either side in a criminal trial, usually only one entomologist was involved in the case. For the first few trials where I testified, only my analysis was presented to the jury. I gave a brief explanation of forensic entomology, told how the evidence had been collected and analyzed, and then presented my interpretations of the evidence. No alternate analysis or interpretation of the evidence was given to the jury. Nowadays, however, it is quite common for both sides in any trial to have their own forensic entomologist if insects are in any way involved.

The direct examination is usually relatively uneventful. I must give the jury a crash course in entomology and ecology before I can present my analyses and have them make any sense. This is like walking a tightrope. Most people do not share my enthusiasm for maggots, and I have to present the subject very carefully to avoid repelling the jurors—and others in the courtroom. Whenever possible, I begin with a slide show, which allows me to explain the life cycles of the different insects clearly and also softens the effect of the slides that will appear later showing the actual corpse and insects. I have discovered that often what I say to a jury is less important than how I say it. I walk a narrow line between avoiding the jargon of my profession, which comes easily to me, and oversimplifying to the point of appearing condescending. And sometimes I unintentionally lapse into terminology that is incomprehensible to anyone but another entomologist.

After I have provided the basic background information, the attorney who has called me leads me through my analysis of the case. This usually includes when I became involved, where I collected the specimens, how they were analyzed, and finally my assessment of the evidence. Usually this is an estimation of the postmortem interval, but sometimes I include other types of interpretations. During this portion of the testimony, slides, photographs, or insect specimens are usually introduced as evidence for the jury. Utmost sensitivity is called for in the presentation of such evidence. The prosecuting attorney I am testifying for wants to make as much of an impact on the members of the jury as possible, while the defense does not want the members of the jury to look at a series of pictures of the victim covered with a mass of feeding maggots. Somewhere a little past the middle of this range is a point at which the horror of what is being presented so overwhelms the individual jury members that they disregard my testimony in disgust. At this point in the trial, the jury and I both get a break while the judge and attorneys argue over the repulsion factors of the various exhibits.

Usually, the actual insect specimens do not evoke any strong reactions and are admitted without much discussion. But I have had a couple of problems in this area of my own making. In one case I had spent a good deal of time examining soil samples associated with a grave. In these samples, I had discovered a new species of mite and I was in the process of describing it and giving it a name when the case came to trial. One of the requirements for describing a new species set forth in the International Code of Zoological Nomenclature is that the specimen on which the description is based, called a holotype, must be deposited in an institution that will make it available for examination by other zoologists. The attorney called me just before I was due in court and, as an afterthought, told me to bring the specimens I had used for my analysis with me. I quickly grabbed all the specimens, including all the specimens of the new species, and took them with me. During the trial, the specimens were entered into evidence and "deposited" as a state's exhibit. Some day I'll get the specimens back and finish the descriptions. I have tried to collect in the same area and similar habitats, but I've never found that species again.

Once I have finished presenting my analysis of the evidence and been examined by the attorney who called me, it is time for the cross examination. Now the opposing attorney, who was so polite during the pretrial interview in my office, must attempt to demonstrate to the judge and jury that I am wrong in my conclusions or a blithering idiot—preferably both. The techniques employed in this effort never cease to amaze me. One popular approach is to pinpoint small pieces of my testimony or report that can be examined in such detail as to leave everyone completely confused. During one trial, I spent well over an hour on the stand using a hand calculator to redo several calculations. The defense attorney had his calculator and I had mine. We entered the same figures and performed the same calculations while the jury became increasingly stupefied and restless. The attorney made so many errors that the judge finally called a halt to the exercise.

Another tactic is to attempt to characterize the entire method I used for analysis as inappropriate for the case under consideration. If my analysis is based on laboratory rearing data, then only field data should be considered appropriate. Of course if I relied on field data, then the laboratory must be the only acceptable source of reliable data. If I used both approaches and arrived at the same conclusion, then I was obviously unsure of myself, since I had had to use both methods. This process sometimes reminds me of a board game where the rules are incidental to the action on the board. Science tends to take a backseat to the legal manipulations, and it frequently requires a major effort to keep the focus of testimony on the problem being addressed.

When, as frequently happens today, both sides have entomologists, they are often acquainted both personally and professionally since there are relatively few forensic entomologists in the United States. In most cases, independent analyses of the evidence result in virtually the same estimates or conclusions. But it is possible in some cases to interpret the evidence differently, and in these cases, the entomologists find themselves testifying for opposite sides. In most of my cases, I have not been allowed to be present during testimony by the opposing entomologist owing to the witness-exclusion rule. This rule is designed to prevent one side from having undue advantage over the other during cross examination, but I'm not sure how well it works. The attorneys usually have me wait in the hallway during other entomologists' testimony and request a break before beginning their cross examination. During this break we review the testimony. In one case, although the other entomologist was not allowed to listen to me testify, his wife was allowed to sit attentively in the courtroom throughout my testimony.

Occasionally, I am allowed to be present as a consultant when another entomologist is giving his or her opinions. Usually this happens not during the trial, but during preliminary hearings dealing with the admissibility of entomological evidence. The conduct of some of these entomologists while presenting their findings has been interesting to watch. In most cases, they seem to do a better job of testifying than they do of giving lectures. Their demeanor while on the stand varies from projecting the persona of a gentle grandfather to radiating complete arrogance. During one of my cases, the opposing entomologist gave a unique explanation of a normal distribution of data: he did an impression of popcorn popping. The judge was impressed, and I can honestly say I've never seen anything like it before or since.

For the most part, however, entomologists' testimony is straightforward and responsible. Only on rare occasions have I heard testimony that is obviously biased and unscientific. In one instance, rather than admit to an obvious error in labeling, an entomologist came up with the explanation that he and "all of his colleagues" had discussed the question at great length and had agreed that this mislabel was an accurate reflection of the truth. Obviously, I was not counted among his colleagues. In fact, I'm not sure who these people were, since no one with whom I have discussed this issue agreed with his interpretations. His testimony finally required so many of these explanations that it was discarded before the case came to trial. He had made an even more basic error in the identifications of the species and developmental stages and this mistake cast considerable doubt on his analyses of the evidence.

When there are two entomologists on opposite sides of a case, one has a much more difficult task than the other. One entomologist must be able, with the assistance of an attorney, to establish the validity of his analysis beyond a "reasonable doubt." The other must simply discover possible areas for raising "reasonable doubt." In a perfect world, this would involve discovering errors in analysis and result in a correct interpretation of the evidence. But this is not always what happens. Even a less than brilliant attorney can make a small discrepancy in data appear to the jury to represent a major error in fact and interpretation. It is easier to sit on the sidelines and throw rocks than to defend one's conclusions from repeated attacks.

In my court appearances, I have also seen some very interesting applications of data. It never ceases to amaze me how the same sets of data can be interpreted differently by the same person at different times for different purposes. Consider, for example, the treatment by an opposing of entomologist of studies I and my graduate students did over a number of years on the life cycles of *Chrysomya megacephala* and *Chrysomya rufifacies* in Hawaii. The results of these studies have been subjected to peer review and have been published in refereed entomological journals. It is also fairly well established by a number of different studies conducted by different researchers that there can be some variation in the duration of developmental stages in different geographic areas. When I used the developmental data I had derived from Hawaiian populations while testifying for the prosecution in a case in Hawaii, the opposing entomologist presented developmental data for those same species from several other geographic areas, including Sri Lanka and South Africa, and claimed that his data showed that my data were incorrect. These studies did give time periods slightly different from those in my data, but he failed to emphasize or even to mention that none of the developmental times given in these studies agreed with each other either. In his testimony, my data were presented as having resulted from the worst possible study ever conducted. I assume the jury was not convinced, since in spite of such attacks on my competence, the suspect was convicted of murder. I later had the opportunity to review the other entomologist's testimony in another case involving the same two species. In that testimony, my times agreed more with his estimate of events and my same published study had magically become a magnificent example of science.

Most of the time, entomological evidence has been reduced to a supporting role by the time a case comes to trial. Establishing the time of death is very important in the beginning, but as the investigation proceeds, other evidence is uncovered and trials tend to focus on that evidence. There are some exceptions, particularly when establishing the time of death has led directly to other evidence. One such case began in November 1996, when the decomposed body of a woman was discovered in a sugar cane field on the island of Kauai. I was traveling on the mainland at the time, so insect specimens were collected, fixed, and preserved in alcohol by Tony Manoukian, a forensic pathologist. On my return to Hawaii, the specimens were shipped to me for analysis. The specimens were in good condition and I was able to identify four species of insects. There were adults of the red-legged ham beetle, *Necrobia rufipes,* larvae of the cheese skipper *Piophila casei,* larvae of the black soldier fly, *Hermetia illucens,* and third instar larvae of a flesh fly in the family Sarcophagidae. On the basis of the species present and the climatic data from the site, I estimated that the body had been decomposing for approximately 34 days before the specimens were collected and preserved. Using this estimate, the police identified a potential suspect. He admitted to having been with the woman on the night I said death had occurred, but said she was alive when he last saw her.

The police obtained a search warrant for the suspect's home. It appeared clean and no traces of blood could be found. Then the police decided to use luminol to try to detect blood stains that might not be visible to the naked eye. Luminol reacts with the iron in the hemoglobin in blood to produce a glowing light. When luminol was sprayed in the bedroom and the lights were dimmed, the outline of a body was clearly visible on the floor beside the bed. The body had been lying on its side and blood had seeped out of it to form the outline. Further treatment showed where the body had been dragged across the floor, through the kitchen, and out the door. Even though the suspect had cleaned the house using bleach and had removed all visible traces of blood, enough residue remained to be detected by the luminol.

During the trial, my analyses became critical because the justification for obtaining the search warrant was almost entirely my estimate of the time of death. If my estimate could be discounted, the search warrant would be considered invalid and all the luminol evidence would be suppressed. In spite of the efforts of the defense attorney to cast doubt on forensic entomology as a discipline, my identifications of the specimens, and the methods used to collect and preserve the specimens, my estimate was accepted by the judge and the damning luminol evidence was admitted.

DURING THE VARIOUS workshops and training sessions I have given for members of law enforcement agencies, I am often asked what these agencies can do to prepare for cases where they find entomological evidence. Usually they expect me to give them a list of equipment to buy and guidelines for collecting and handling specimens. Instead, I suggest that these agencies go to their local college or university and "collect" an entomologist. In the end, this will serve them better than any set of instructions I can provide.

An entomologist at a college or university in the vicinity of the crime scene will have the benefit of knowledge of the local insect fauna. This will increase the chances of a representative sample's being collected and properly preserved for analysis. I have also suggested that the entomologist be "collected" before he is needed, so that he will have time to decide if he wants to become involved and to what extent. Not everyone is willing to participate in a criminal investigation, and for that matter, not everyone who wants to be involved in an investigation should be allowed to participate.

As the acceptance of entomological evidence grew, a number of agencies took my advice and did contact local entomologists for assistance and tried to recruit them as consultants. Keep in mind that most potential forensic entomologists are university professors, people who are used to talking to often skeptical students and colleagues. When they are approached by someone in a position of authority, usually brandishing a badge, who wants their help and is willing to listen to their opinions, their behavior patterns begin to resemble those of a golden retriever—overly enthusiastic and eager to please. I make this comparison not to offend either entomologists or dog owners, but because that is the behavior I have observed in both my colleagues and my dog, Ginger, a golden retriever–Irish setter mix. Give her attention and a dog biscuit and she'll do almost anything. Entomologists are often as ardent as Ginger (except for the part about the biscuit), and must be introduced to criminal investigations with some care.

Most beginning forensic entomologists do not have much experience in dealing with the dead, and their reactions to a corpse will vary a lot. I have one colleague who has chosen to work in forensic entomology even though he becomes violently ill at the sight of a body and cannot bring himself to go into the morgue, let alone visit a crime scene while the body is still present. Any collections he uses in his estimates must be made by others, and they are therefore only as reliable as collections made by any third party. Another colleague is more than willing to go to any crime scene and often tries to participate in areas of the investigation well outside the limits of his expertise. Like most of my colleagues, I fall somewhere in between in my responses to corpses and my involvement in investigations. Over the years I have developed several techniques for dealing with death, and although I am certainly not comfortable with death, particularly violent death, I am now usually able to manage my emotions. It is essential that before an entomologist, or anyone else for that matter, becomes involved in a criminal investigation he or she be made aware of what to expect.

As more entomologists became aware of forensic entomology as a new and exciting area, many decided to join the ranks. Frequently they made this decision and initiated contact with their local law enforcement agencies without being fully aware of the possible consequences. Some did not seem to realize that once they began an investigation, it would not be easy simply to say, "I'm tired of this and I want to go home now." If the entomologist makes collections from the body, he has taken evidence from the corpse. This evidence is part of the investigation and must be handled appropriately. If he continues, completes an analysis, and files a report, he has further involved himself in the investigation. Each of these steps leads to deeper levels of involvement and responsibility. In my career, I have encountered many entomologists who began a case with naive enthusiasm, not realizing that they would ultimately be required to present their results during a trial and in the presence of the accused. They began with a sense of excitement and willingly offered opinions and speculations that were not supported by the evidence. This is a very easy trap to fall into. By the time I entered these cases, the entomologists had become painfully aware of their mistakes and were facing the prospect of having to testify. In all of my forensic entomology investigations, I am keenly aware that I may have to face someone whose life can be radically changed—even ended—by my testimony. I do not take this responsibility lightly and am careful to analyze and reanalyze my data and then scrupulously qualify my estimates to prevent misinterpretation of the results. When someone's life is at stake, all the enthusiasm in the world is a poor substitute for accuracy.

OVER THE YEARS a number of my colleagues, both those actively involved in forensic entomology and those simply watching from the sidelines, have expressed dismay over the interactions between entomology and the legal system. There are definite problems to be overcome and I certainly do not claim to have all the answers. In fact, I don't even have all the questions yet. Whenever science and our legal system collide, problems will arise. We will continue our research and eventually, I hope, develop a system that minimizes conflicts between entomology and the courts. Any system will of course be only as good as those participating in it and, regrettably, as has happened in other disciplines, there will probably always be some entomologists who, for a fee, are willing to bias their testimony. Several years ago during meetings of the American Academy of Forensic Sciences, Joseph Davis, the chief medical examiner for Dade County, Florida, expressed fears that forensic entomologists would become the newest in a series of traveling witnesses for hire. As much as I would have liked to jump up and contradict him, I knew better. I had already seen indications that this was happening.

Faced with this situation, my colleagues and I could either wait for the problem of "hired guns" to reach the point where entomological evidence was again ignored or take action to introduce standards for ourselves. Beginning in 1993, the topic for the annual breakfast of the Dirty Dozen became the need for a certification board. Some of us thought that the only logical place for our board would be in the Entomological Society of America. Others were more comfortable with an independent board, which would, they hoped, be recognized by both the Entomological Society of America and the American Academy of Forensic Sciences. Since many more people than the members of our original group were now involved in forensic entomology, we decided to survey as many people as we could who were interested in the field. After two surveys it became apparent that most people would not respond to surveys. But those who did respond tended to favor an affiliation with both organizations.

Ultimately, we decided on an independent board. The core group involved in creating this board consisted of Paul Catts, Valerie Cervenka, Robert Hall, Neal Haskell, K. C. Kim, Wayne Lord, Ken Schoenly, Ted Suman, Jeff Wells, and me. For the next couple of years, we passed around various drafts of protocols, by-laws, and suggested reading lists. Finally, in 1996, we were able to officially incorporate in the state of Nevada as the American Board of Forensic Entomology. Our first board of directors was elected in February 1996 during the annual meeting of the American Academy of Forensic Sciences in Nashville, and the members were Gail Anderson, Paul Catts, Robert Hall, Wayne Lord, and me; I was the chair. Our incorporation became official on April 2, 1996. Paul Catts died several days later while playing lacrosse and was replaced by Neal Haskell. We have already issued our first certifications of forensic entomologists. There will probably still be conflicts between the forensic entomologists and the legal system, but we have begun to establish a set of standards to resolve them.

12

Spreading the Word

As forensic entomology became more well known, interest increased in groups other than medical examiners and police. This interest led to inclusion of presentations on forensic entomology at conferences dealing primarily with other areas of forensic science. Among these were the international symposia series hosted by the Federal Bureau of Investigation at the FBI Academy in Quantico, Virginia.

I first arrived at the FBI Academy on June 23, 1990, to give a presentation on the effects of drugs on maggot growth during the International Symposium on Mass Disasters and Crime Scene Reconstruction. This symposium was my first exposure to the FBI Academy and began an association that has proven highly productive over the years. I was selected to participate largely because of the efforts of Wayne Lord. At that time Wayne was a special agent assigned to the New Haven office and, in addition to assisting in the organization of the symposium, was giving the presentation on forensic entomology. I was not sure what to expect from the symposium or what to expect from the Academy. On the bus from the airport to Quantico I began to realize that I was entering a different world. My Levis, shirt, and sandals were a definite departure from the conservative attire of the rest of the participants. On my arrival, I was handed a flier detailing the Academy's dress code and I realized that I was violating several of its sections. I made efforts to adhere to the code but retained my beard, my shaggy hair, and the earring in my left ear. A couple of the senior agents never could get used to my earring, and one of them spent almost the entire week having conversations with the diamond stud in my ear.

Considering its mission, Quantico was a strangely comforting facility. Everyone was pleasant, even when practicing how to handle hostage situations or how to apprehend "suspects" in the dorm hallways in the wee hours. The doors are not locked, so people don't feel caged in. The entire complex is connected by tunnels and enclosed walkways and every part of it is air-conditioned, so the temperature inside is always comfortable, even in the sweltering Virginia summer. The walkways form a giant maze that might have been constructed by a psychologist who got tired of working with rats and decided to experiment on people. I found myself expecting to see someone with a stopwatch appear around a corner or to come upon a lever to be pressed for a treat. But all things considered, it was a very comfortable place in which to discuss the atrocities committed in the outside world.

The choice of recreational activities was, however, limited. The main gathering area in the evenings was a combination snack bar–beer hall called the Board Room. It closed at 10:30 P.M., along with everything else. Since it was late June, warm outside, and there was a parking lot with powerful lights, Wayne and I began to behave like entomologists. We went out into the parking lot and started collecting the insects attracted to the lights. Only 10 minutes later security guards appeared and asked what we were doing; after a lengthy discussion they decided we were harmless and let us continue. The next evening, we were joined by a couple of others and by the third night we were a pack, combing the parking lot for dazed insects that had flown into the lights. Even the security guards began to get interested.

I HAVE PARTICIPATED in many FBI-sponsored symposia covering a variety of forensic topics, from toxicology through trace evidence to arson investigations. For the past several years, I have been involved in an annual training session for FBI agents on the detection and recovery of human remains in outdoor settings. As part of this session, as I mentioned earlier, I have a series of pigs placed in commonly encountered situations—usually dumped on the ground, shallowly buried, and hanging. Over the course of the week, the participants see the changes in the pigs and gain hands-on experience in collecting insects from the carcasses.

As the week progresses, the agents use different tools and, not unnaturally, invent nicknames for them. During the session in May 1996, the geophysicist's instruments became the "yellow thingy" and the "red thingy," and the cadaver dog became the "brown furry thing." When the director of the FBI visited the class on the last day and asked participants what should be provided for the individual teams, he looked disconcerted when asked for a "couple of the red thingys and a brown furry thing."

My pigs and I did not escape completely unscathed. Some level of decorum was maintained until the night before the grave excavation practicum. But when I visited the pigs early the next morning before the exercise, I was amazed to see how well they had dressed for the occasion. Each pig had on a pair of glasses with a plastic nose, neckties were in place, and one was wearing a bib. Of course, no one in the class would admit having been anywhere near the pigs, but there were three female agents who seemed suspiciously enthusiastic about posing with the well-dressed pigs.

ACCEPTANCE OF FORENSIC entomology by the FBI and its inclusion in their symposia and training programs constituted a major step forward. Forensic entomology was being taken seriously and gaining wider acceptance within both the legal and the forensic communities. This was the result of continuing efforts by members of the Dirty Dozen to establish forensic entomology as a distinct discipline through participation in various professional organizations.

My first excursion into this educational effort came in 1985. I had been working with the medical examiner in Honolulu, Charlie Odom, for a couple of years and he had become convinced that there was a use for entomology. He suggested that we might present a couple of cases at the annual meeting of the National Association of Medical Examiners. That year the meeting was to be held during September in Memphis, Tennessee. It sounded like a good idea, and I agreed to go. I got to Memphis early in the morning and went directly to the meeting site at the Peabody Hotel, arriving just in time to witness a spectacle—the morning "duck break." I later learned that this was an event unique to the Peabody Hotel and had occurred every day for over 50 years. All activities in the hotel seemed to stop as the Peabody ducks made their entrance. I was suffering from severe jet lag, and the sight of a bunch of people cheering on a group of ducks marching across a red carpet laid from an elevator to a fountain in the center of the hotel lobby—all this to the accompaniment of a Sousa march—was almost enough to send me back to the airport to catch the next flight to Honolulu. But the taxi had left and just then Charlie found me, so I stayed. I even was amused when I discovered that there was a "Duck Palace" on the roof of the hotel.

The meeting itself was a definite departure from what I had encountered in entomological circles. At the average entomological meeting, most of the attendees have chairs. The above-average entomological meeting provides chairs for everyone and pitchers of water at the rear of the hall. Here there was a series of tables with writing materials neatly arrayed and pitchers of ice water at each place. The presentations were somewhat less formal than what I was used to but were well done. In our presentation Charlie gave the pathologist's view of the cases and I provided the insect information. In picking the slides, I had carefully selected those I felt would be most appropriate for a group of medical examiners—the ones showing corpses. I was completely wrong. Pathologists are all too familiar with dead bodies. Insects, by contrast, are fascinating. Since then, I never show a series of slides of just bodies in presentations to pathologists or show only insect slides at entomological meetings. Pathologists want to see the insects and entomologists want to see dead bodies. Mixed groups require mixed slides, and lawyers are amused by almost anything.

The meeting of the National Association of Medical Examiners in Memphis was a turning point in my career. Until then, I had regarded my involvement in forensic entomology as a sideline. After those meetings, my outlook changed and I focused my efforts. I had discovered a group of dedicated, interesting people intensely involved in finding solutions to problems I had only begun to explore. Among them was Bill Rodriguez, an anthropologist with whom I have since worked closely on several projects. My initial impulse was to join the Association and become more involved. Charlie Odom put a temporary damper on my enthusiasm by pointing out that I was not qualified for any level of membership in the National Association of Medical Examiners. Full membership is restricted to medical examiners, physicians, and coroners who are directly involved in the investigation of death. Affiliate membership is restricted to people involved in death investigations on or approaching a full-time basis who are affiliated with a coroner or medical examiner's office. There seemed to be no provision for university professors interested in insects. Charlie did suggest another organization that, he thought, might be more appropriate: the American Academy of Forensic Sciences.

AFTER MY RETURN to Hawaii from the Memphis meeting, I obtained the application forms for the American Academy of Forensic Sciences. There are several sections within the Academy and I had to select one to put on my application. I chose the Pathology/Biology Section. Membership in the Pathology/Biology Section is open to people holding M.D., D.O., or Ph.D. degrees, and I qualified with my Ph.D—it was only later that I discovered how rare Ph.D.'s were in the Pathology/Biology Section at that time. I was voted into provisional membership during the 1986 meeting and attended my first meeting in San Diego in 1987.

The American Academy of Forensic Sciences meetings make the National Association of Medical Examiners meetings look reserved. Virtually everyone who had been at the National Association of Medical Examiners meetings was also in the Pathology/Biology Section of the Academy and there were nine other sections, each with at least as many members. During the San Diego meeting I spent a good deal of time feeling somewhat lost because of the immense number of people attending and the physical layout of the Town and Country Hotel, the same hotel where I had first been exposed to forensic entomology at a meeting of the Entomological Society of America. This time, I was able to stay in the hotel and not in another hotel across the freeway. But there were so many different presentations in so many different areas that I was continually moving from one meeting room to another and still feeling that I was missing some presentation I should be attending. It took a couple of years for me to feel comfortable at Academy meetings.

But I quickly learned that there are two "must-attend" events: The Last Word Society and Bring Your Own Slides. The Last Word Society involves a series of presentations in which people reanalyze cases or events of historical interest using techniques not available at the time they occurred. Bring Your Own Slides is a session that began with a small group of Academy members meeting in a hotel room to discuss unusual cases. It has now evolved into one of the best attended events during the meetings and requires a very large ballroom. Refereed by Michael Baden, it is a session during which members present unusual-to-fantastic cases, frequently with great humor. At these sessions, it has become traditional to have a bottle of Jack Daniels available. During the Seattle meeting, the local regulations required that the bottle be under the care of a bartender. I arrived a few minutes after the start of the presentations, and walked up to the portable bar at the rear of the room. I did not see a bartender, but as I stood there, I heard a voice from under the bar ask what I wanted. I peered over the bar and saw the bartender and his assistant crouched behind the bar looking at the wall. I ordered a beer and it was promptly handed to me, and I paid and received my change, all without having either man look at me. Apparently the bartenders had thought they would be working at a flower show that was scheduled for later that week, and instead of the evening of orchids they were expecting, they were being forced to listen to a pathologist from England talking about the fatal results of a do-it-yourself sex-change operation.

My early presentations at these meetings were viewed as sideshows. The potential use for my data was recognized, but most people still regarded the bugs as a novelty rather than a routine part of an investigation. For the meeting in Las Vegas in 1989, Bill Rodriguez approached Wayne Lord and me with a proposal for a workshop on the recovery of decomposed bodies. He had lined up a medical examiner, Larry Tate, and another anthropologist, Bill Haglund. Wayne and I would round out the group as the entomologists. As Bill explained the idea to me, we would have a relatively small group of 30 to 40 people and we would sit around a couple of tables for a hands-on workshop dealing with the techniques of anthropology, entomology, and pathology that are commonly used to recover evidence from a decomposing body. The idea sounded good to me and I enthusiastically agreed. As the preregistration process continued, the response was better than expected and the Academy office called Bill to ask if a "few more" could register. This happened a couple of times and finally Bill agreed to take the lid off of registration. We ended up with 175 people. We had to abandon the idea of a hands-on workshop with a pass-the-specimen approach. So we decided to opt for a more formal lecture followed by a series of demonstrations on tables placed around the room.

One of the demonstrations we thought would be essential was an exercise in recovering maggots and pupae from soil placed in a large, shallow box. But since we would be indoors in a hotel ballroom, we decided we should use sawdust instead of dirt. The plan was for each participant to find a pupa in a box and put it into a glass vial we would supply. Each participant would then be able to see the adult fly emerge from the pupal case when it completed its development during the course of the meetings. The maggots were ordered from a biological supply company and we told the company to ship them to the hotel to arrive the day before our workshop began, but the package arrived several days early. Since it was February and cold in Las Vegas and the box was clearly labeled "fragile, live material," the hotel staff decided to put it in a warm place until we arrived. We had ordered 1,000 maggots. When Bill and I picked up the box the morning before the workshop, we thought we heard a couple of adult flies. We went to the ballroom where the workshop was to be held and opened the box. Much to our dismay and the amazement of the hotel workers still setting up for the workshop, a cloud of adult flies emerged and began exploring the room. There were hundreds of flies in that and several other rooms for the entire meeting, and the workshop participants had to content themselves with finding empty pupal cases in the sawdust.

In later workshops, we have had better luck putting the pupae into glass vials and then passing out the vials to the participants. That way the adults emerge inside the vials instead of all over the meeting rooms. Much to my astonishment, people often become quite attached to their pupae. I still am startled when someone comes up to me—usually someone I don't remember—late in the evening and in a conspiratorial tone says something like, "Mine isn't out yet."

Through the years, more and more forensic entomologists have joined the Academy, both in the Pathology/Biology Section and in the General Section. Now we are an accepted part of the Academy rather than a fringe contingent. Our level of participation in the affairs of the Academy has also increased with time. Wayne Lord and I were the first entomologists to become Fellows in the Academy, and I was elected secretary of the Pathology/Biology Section in 1995 and chair in 1997, the first person to be chosen an officer of the Pathology/Biology Section who was not a forensic pathologist.

IN 1985, THE forensic community was not alone in regarding forensic entomology as a fringe activity. Many members of the entomological community were skeptical about, if not downright distrustful of, forensic entomology, although most of them were fascinated by the subject matter. The reactions to presentations on forensic entomology at national and international meetings became embarrassing. At one of the first presentations I gave for a national meeting of the Entomological Society of America, my paper was preceded by an admirable presentation on stable flies and followed by one on fleas. At the beginning of the paper on stable flies, the room was about half full. The last half of that presentation was lost in the noise of the door opening and closing as the room began to fill. By the time I began my presentation, there was standing room only. When I finished, most of the audience got up and left. The presenter of the flea paper, which was excellent, had a small audience that filled only about a third of the room. The next year, both presenters demanded that their papers not be scheduled in the same session or even on the same day as mine.

Realizing that we needed support in our efforts to establish forensic entomology as a distinct discipline, some of my colleagues and I began to organize symposia and workshops for entomologists both in the Entomological Society of America and the Society of Vector Ecology. One of the first symposia I organized was for the annual meetings of the Society of Vector Ecology held at the University of California at Irvine in 1985. I assembled a series of speakers who would cover both entomological subjects and topics related to the medical and legal concerns forensic entomologists were increasingly encountering. Kate Tullis, one of my graduate students, gave a presentation on the results of her research in the rain forests of Oahu on pig decomposition; Paul Catts spoke on future possibilities for research and forensic applications; a district attorney addressed the legal problems; and Warren Lovell, a forensic pathologist, gave some insights into pathology and mass disasters.

The symposium was held in an "expandable" room located in the student union. The organizers had underestimated the attendance, and the first part of Kate Tullis's talk was lost as we opened partitions to accommodate an overflow crowd. After everyone was seated, it became apparent that the sound system was operating in only the front part of the hall. Quickly the organizers turned a few dials and flipped a couple of switches, the sound came on, and things proceeded nicely. During Warren Lovell's presentation, I took a quick walk outside to get a little fresh air. Just beyond the entrance to the room, there was a walk-up post office and since it was close to Christmas mailing deadlines, people were standing in line to mail their parcels. As I walked out, I saw people standing in a circle looking up at a speaker mounted in the ceiling. Before the symposium began, the speaker had been sending forth Christmas tunes. But now I could hear Warren giving an account of a train crash in which he had found more legs than could be accounted for by the bodies available. Apparently when the organizers adjusted the sound system, they unintentionally sent the audio portion of the symposium into all of the campus center, including the bookstore, the cafeteria, and ski shop. No one complained and, throughout the center, there were groups of people standing around speakers listening to the symposium.

The symposia presented during the annual meetings were successful, and forensic entomology was becoming a "hot" topic, so, in 1992, officials of the Entomological Society of America asked me to organize a one-day workshop to follow the annual meetings in Baltimore. This was to be their first effort at providing continuing education associated with their annual meetings. I agreed, with the provision that the presenters would be the same people I was working with on the workshops for the American Academy of Forensic Sciences. I contacted some of the usual perpetrators—Bill Rodriguez, Wayne Lord, and Ed McDonough—and they were willing to participate. I then added Ted Suman for his expertise on the local insect fauna. Ted was then relatively new to forensic entomology, and I had not seen him for several years, since we were both at the B. P. Bishop Museum in Honolulu, where he was working on spiders and I was working on mosquitoes.

The workshop went quite well. For once, all the presenters managed to get there and give their own presentations. Over 180 people attended, and the evaluations were so good enough and the response so great that the Society wanted us to give the same workshop at the annual meeting the next year. Our previous experiences with workshops at the American Academy of Forensic Sciences suggested that giving one every other year would work better. But the Society decided to go ahead and give the same workshop, using other presenters, for the next 3 years. Each year the attendance dropped and the workshop is no longer given. There are only so many people who will take a workshop on any given topic, and almost none who will take it annually. In contrast, by adhering to an alternate-year schedule for our workshops at the Academy meetings we have maintained a fairly consistent level of attendance.

IN SPITE OF the successes of these symposia and workshops, the general response by the entomological community to the field was puzzling to me and the rest of the Dirty Dozen. On the one hand, whenever we gave a presentation, there was standing room only. On the other hand, the entomological societies seemed reluctant to recognize forensic entomology as a separate discipline. We realized early on that some form of recognition by a board or other regulatory body was necessary for us to function effectively in the court system. In the courts, as I knew well from my own time testifying, most people providing expert testimony were certified by one board or another. The legal system is geared to recognize expert witnesses with board certifications; academic degrees are not sufficient.

First, we looked to the American Registry of Professional Entomologists for certification and then later to the Entomological Society of America. Both organizations responded that we could become certified in medical and veterinary entomology but that no special certification in forensic entomology would be offered. As our efforts continued, forensic entomology was listed as an area of specialization in the Society, but no certification specifically in forensic entomology became available.

Although I had been told that I was not qualified for any level of membership in the National Association of Medical Examiners, I still thought that some affiliation with that organization would help promote forensic entomology as a distinct discipline. During my first meeting of the American Academy of Forensic Sciences, I spent some time discussing my feelings with George Gantner, the medical examiner for the City and County of St. Louis. He expressed the same reservations as Charlie Odom had earlier, but he was quite supportive of an increased role for entomology in death investigations. After that meeting, I managed, by somewhat devious means, to obtain and submit an application for membership. Charlie Odom told me that my application elicited a lot of discussion when it was considered during the next annual meeting. To my surprise, I was admitted as an affiliate member. Not everybody agreed with the decision, I suspect, because a couple of the members still refer to me as the Association's illegal member. To date, other forensic entomologists have not been as successful in joining this organization, and I am still the only forensic entomologist listed in the National Association of Medical Examiners *Directory of Members.*

My participation in the National Association of Medical Examiners has not been as great as in the American Academy of Forensic Sciences. One reason is that I am tenured at the University of Hawaii at Manoa in the College of Tropical Agriculture and Human Resources, and it is difficult to convince the administration that my attendance at the National Association of Medical Examiners meetings is directly related to the future of agriculture in Hawaii. Consequently, I usually manage to attend only the interim meetings, which are held during the meetings of the American Academy of Forensic Sciences. The national meetings of the National Association of Medical Examiners are usually held in places affording a certain amount of privacy. Given the materials presented, it is best that the general public not be allowed to wander into the sessions.

AS THE INTEREST in entomology increased among the forensic community, there was a corresponding increase in interest in the forensic sciences among students in various entomology programs. Students today are quick to note a new and developing discipline that may at some point offer them employment. They are much more job focused than the students of the 1960s and do not take as many detours as students did during my undergraduate and graduate days. They concentrate on courses and programs that will lead to jobs, and anything not directly related to the targeted position is either bypassed or endured grudgingly.

Applications from students wanting to work in my laboratory have increased dramatically, and for the past 10 years the number of applications has greatly exceeded the number of available assistantships. Moreover, the applicants have come from an unusual assortment of undergraduate majors, including fashion design, business, mechanical engineering, and music. This variety of backgrounds has made for a diverse and interesting set of students. I have enjoyed working with them all, and the number of publications resulting from their research shows that we have had an exceptionally productive series of interactions.

DURING THE LATE 1980s and early 1990s, I traveled to many universities to give presentations and workshops on forensic entomology. These were pleasant and stimulating excursions, if sometimes quite tiring. The organizers of most of these trips seemed to think that I needed to have activities scheduled for virtually every minute of my stay.

One particularly memorable trip was to a university in North Carolina, where I was to deliver one of several "memorial lectures" as well as conduct a workshop. Owing to some changes in my teaching schedule, this rapidly became the trip from Hell. I caught a flight that left Honolulu at 10:00 P.M. on Thursday night. After two changes of planes, I arrived in Raleigh at 11:00 A.M. Friday morning. I managed to get 2 hours of sleep on a couch in one of the student organizer's homes and gave my lecture at 3:00 P.M. This was followed by meetings with a couple of interested students, a reception, happy hour at the local pub, and dinner at a brew house restaurant. I did not get to my hotel until almost midnight. Unfortunately for me, my visit coincided with homecoming weekend and the Big Game. The game was on Saturday, so the alumni played football in the hotel hallways all night and all day Saturday until the game began. After an unrestful night, I had breakfast on Saturday morning with students at 8:00 A.M., and then I conducted a workshop that ran until noon. I spent the afternoon being driven around the vicinity to see the sights and then went to a party on Saturday night, followed by a flight back to Honolulu that left at 6:30 A.M. the next morning and arrived in Hawaii at 9:30 P.M. I taught at 8:00 A.M. the next morning. Luckily, I had managed to get some sleep on Saturday night, primarily because the home team lost and the alumni were very subdued.

In a moment of mental lapse, I once agreed to give a seminar for the Biology Department at the University of North Dakota in February. I say mental lapse because I am not accustomed to cold weather and forgot about the time of year. I look for extra blankets at temperatures below 70°F. Bill Wrenn, a good friend from my days of studying trombiculid mites, suggested that I would enjoy seeing his laboratory and giving a seminar. I agreed to come to Grand Forks after the annual meetings of the American Academy of Forensic Sciences in New Orleans. I packed for New Orleans in February but not for the weather in Grand Forks. I had never seen that much snow on such a flat landscape in my life. While I shivered in my cotton aloha shirt, I saw people walking around wearing shorts and T-shirts! The seminar was better attended than I had anticipated and even the local TV stations came to cover the event. Since I was going to present some rather graphic slides, I warned the reporters beforehand, and they assured me that they would certainly not use any of *those* shots in their coverage. I guess I was a little naive. After the seminar, Bill and I were having dinner when I glanced up at the TV screen in the bar area of the restaurant. There I was silhouetted in front of a maggot-infested skull. I was slightly out of focus.

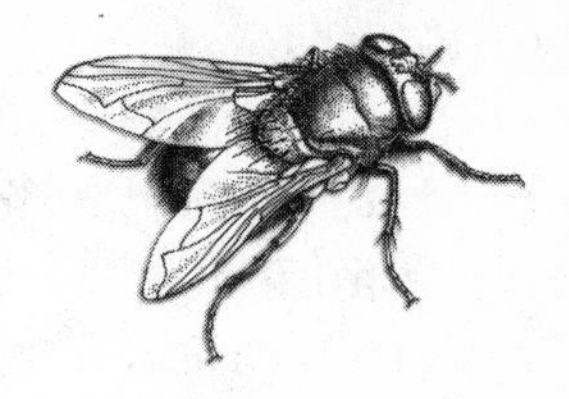

Epilogue: Summing Up

Nowadays, new paths of investigation in forensic entomology are opening up almost daily. Recently, some of my colleagues who were working on an unusual rape case analyzed the blood in the gut of a pubic louse for human DNA content. Students in my laboratory are planning to repeat this process using the gut contents of bed bugs. The implications of such analyses are far-reaching, and it may not be too long before DNA data derived from insects provides vital evidence linking suspects to crimes, especially rape and murder.

Old problems are being revisited. My students and I are investigating differences in decomposition patterns related to the location of the body. A corpse that has been partially immersed in salt water has always presented me and other scientists with entomological difficulties. So one of my graduate students examined the differences between a corpse deposited in an intertidal area and one dumped on dry land. This proved to be the most difficult study I've ever been associated with. The physical logistics were relatively simple, but not the process of obtaining the permits required to place a dead pig in a tidal area. I found that I had to obtain permits or clearances from the U.S. Environmental Protection Agency, the U.S. Fish and Wildlife Service, the National Marine Fisheries Service, the U.S. Coast Guard, the Army Corps of Engineers, Hawaii's Department of Land and Natural Resources, the Clean Water Branch of the State of Hawaii's Department of Health, and the Department of Business, Economic Development, and Tourism of Hawaii's Office of Planning. At one time there was even some concern that a dead pig on the shore might be a hazard to navigation. I was tempted at that point to suggest placing a flashing red or green light on the pig's nose, or even to ask the student involved to change his thesis topic from "Decomposition Patterns in Shoreline Habitats in Hawaii" to "How to Get a Permit to Conduct Research in the Water." How "tourism" could be related to this study is a mystery to me, although it is true that once the pig was in place, it became a tourist attraction of sorts. About half a mile away from the study site on Coconut Island in Kaneohe Bay, there is a tour operator based on a barge who provides kayaks and jet skies for Japanese tourists. The exclosure cages on the shore became a magnet for these tourists. I finally had to get some Crime Scene tape from the Honolulu Police Department and rope the study area off.

On dry land, I have been doing more research on bodies found hanging. Although I already have a lot of data about decomposition in bodies that are hanged, there is much more to be learned. For 3 months I had a Ph.D. candidate from Brazil working in my laboratory whose primary research dealt with drug detection and drug interactions. She wanted to learn the techniques I have developed for conducting decomposition studies. So, along with another Ph.D. candidate from Alexandria University in Egypt who spent a year in my laboratory, she conducted one of the arid habitat hanging studies inside Diamond Head Crater. The field portion of the study is completed and we are now analyzing the data.

The questions posed by forensic entomology may seem new. But in reality, many are refinements of old questions. What's new is that the answers can now be made more precise and more detailed. Advances in technology have allowed me and other forensic entomologists to be more accurate in our investigations and to consider evidence that was previously unavailable. At the same time, I find myself looking again at old questions regarding human nature. Some things I am certain I will never be able to comprehend. Recently there was a case from the island of Kauai in which a woman was beaten severely, was bound with tape, had a plastic bag taped over her head, and was then carried from her house. All this happened in front of at least one witness. Her body was discovered the next day in a shallow grave a short distance from the house. It was easy for me to find entomological evidence that was consistent with the eyewitness accounts, though it really was not needed to establish the time of death. What I could not easily do was understand why none of the eyewitnesses made any attempt to intervene or call the police.

AS FORENSIC ENTOMOLOGY has changed, I also have changed. When I first entered the field over 15 years ago, I was an entomologist and acarologist. My concerns were insects and mites and their relationships with each other, and only occasionally their impact on humans as agents or vectors for diseases or as the cause of damage to homes and products. Since that time, I have developed, or metamorphosed if you will, into something different. I am no longer concerned with just the insects and what they do to each other. I have entered into an area where insects, crime, and our criminal justice system intersect with the very darkest aspects of human nature.

There is now an immediacy to my efforts that is often missing from purely academic pursuits. This new area of forensic entomology moves very rapidly. It must. When a murder has been committed, the police do not have the luxury of waiting several years to take action. They do not have time for grant proposals to be written, submitted, and funded to facilitate research to be conducted for a specific case. Information is needed immediately and investigations tend to move quickly to a conclusion. Consequently, when I am confronted with an unusual set of circumstances, much of my research has to be conducted initially in a "quick and dirty" manner to provide immediate direction for the investigation. Once this has been done, I can conduct more controlled studies, with, I hope, some form of grant support. Indeed, situations that I have first encountered during a death investigation have suggested some of my most interesting long-term research projects.

While I was involved in research on the mites, I was rarely, if ever, considered newsworthy. If fact, between 1977 and 1993 I was asked to speak in public about my specialty only twice. After I took up forensic entomology and testified in a few trials, I was suddenly in great demand. The public is fascinated by murders, and I have more requests to speak than I can possibly fill and still teach and conduct research at the university.

The public's interest in forensic science has been fueled in recent years by the news media's reporting of investigations of homicides and other crimes. As I left the court after my first experience testifying, I was approached by a local reporter who wanted to write an article on what I had done in the case. I was a little reluctant to be interviewed, but eventually I did consent and the result was my picture peering out from the evening newspaper. The reporter had done a good job of describing my work, possibly too good a job. For the next couple of days, I was aware of people looking at me a little strangely. Since then, I have been written about in a number of popular publications and have appeared in several television documentaries. These are always a challenge. The directors want realism and often would prefer me somehow to schedule the discovery of a body during the time they plan to be filming. Even if I could do so, which of course I can't, they would not be allowed to film the crime scene. Usually they settle for a reconstruction of a crime, as in the case in which I wrapped a dead pig in a blanket to replicate the disposal of the body.

Another approach is to film me working in the laboratory and then do an interview at the site of a previous crime. I recently did one such interview that was memorable for its lack of light. The director wanted to convey an "eerie" feeling, and filmed most of the interview in a darkened room with only red and blue lights illuminating me and my specimens. The climax was the filming of a crime scene along the old Pali Highway—at 9:00 P.M., in keeping with the "dark" nature of the subject. There we were, an eight-person film crew and me, on the side of a road in a rain forest, where the body had been left only poorly concealed from sight, illuminated by car headlights, with the camera showing only dangling vines and tree trunks along the roadside. When I was working on the actual case, I had often wondered why no one had noticed the activities of the people dumping the body. The area appeared remote from the angle of the filming, but there were several houses on the opposite side of the road and less than 15 yards from it. We were at the site for over 45 minutes, with flashing red and blue lights, and my Harley going up and down the road, yet no one seemed to take any notice. I suppose the suspects in this case did not have to hurry at all. In fact, they probably had all the time in the world. Had they known this, I'm certain they would have done a better job of concealing the body than they did.

One problem I had not anticipated with regard to the publicity my work has received through these television appearances is communications from people who want more information. Every time one of these television segments is aired, I receive lots of letters and e-mails. Most are of a general nature, usually from students wanting more details and occasionally looking for graduate schools. I answer these as best I can and try to provide references that will lead these inquirers to more information.

Other requests are disturbing and come very close to asking me to assist the writer in designing a perfect crime. One e-mail inquirer was very persistent in seeking my advice on how to dispose of a body in Hawaii and how to alter the entomological evidence to prevent accurate estimation of the postmortem interval. Another sought my help in setting up a system to detect serial killers and track them down to dispose of them before they were able to kill. This man claimed to have personal experience with serial killing through a relative and believed he knew more about serial killers than the FBI and other law enforcement agencies. Usually I don't answer these types of inquiries at all. In this case, I made an exception, but the inquirer disappeared into the Internet. I still wonder about him occasionally. Probably he was harmless and was merely indulging in some game-playing. But he could have been a serial killer trying to improve his techniques.

WHEN PEOPLE FIND out what I "do for a living" they usually have two questions. The first is, "How can you get used to all that death and violence?" The second is, "What has been your most interesting case?" I get other questions about smells, sleeping at night, worry about revenge by a murderer, or how I came to get into this field, but these two are by far the most frequent. I don't ever plan on becoming so used to death and violence that I am no longer affected by it. No one should ever be that impervious. I deal with it by being as detached as I can and by falling back on gallows humor to relieve the tension, like everybody else who has to deal with violent death. What my most interesting case is I still don't know. Possibly when I quit I'll have an answer. But even then, the choice will be hard. For now, I'm still waiting for that case. Each one is different and presents new challenges.

Bibliography

Acknowledgments

Index

Bibliography

Bergeret, M. 1855. Infanticide, momification du cadavre. Découverte du cadavre d'un enfant nouveau-né dans une cheminée ouil sétait momifié. Determination de l'époque de la naissance par la présence de nymphes et de larves d'insectes dans le cadavre et par l'étude de leurs métamorphoses. *Annals of Hygiene and Legal Medicine* 4:442–452.

Beyer, C. J., W. F. Enos, and M. Stajic. 1980. Drug identification through analysis of maggots. *Journal of Forensic Sciences* 25: 411–412.

Blackith, R. E., and R. M. Blackith. 1990. Insect infestations in small corpses. *Journal of Natural History* 24:699–709.

Bornemissza, G. F. 1956. An analysis of arthropod succession in carrion and the effect of its decomposition on the soil fauna. *Australian Journal of Zoology* 5:1–12.

Catts, E. P., and M. L. Goff. 1992. Forensic entomology in criminal investigations. *Annual Review of Entomology* 37:253–72.

Catts, E. P., and N. H. Haskell, eds. 1990. *Entomology and death: A procedural guide.* Joyce's Print Shop, Clemson, S.C.

Coe, M. 1979. The decomposition of elephant carcasses in the Tsavo (East) National Park, Kenya. *Journal of Arid Environments* 1:71–86.

Cornaby, B. W. 1974. An analysis of arthropod succession in carrion and the effect of its decomposition on soil fauna. *Australian Journal of Zoology* 5:1–12.

Dreher, G. C. 1933. Maggots: Their experimental uses in dentistry. *Dental Survey* 9:26–38.

Early, M., and M. L. Goff. 1986. Arthropod succession patterns in exposed carrion on the island of Oahu, Hawaiian Islands, USA. *Journal of Medical Entomology* 23:520–531.

Erzinçlioglu, Z. 1985. Few flies on forensic entomologists. *New Scientist* May:15–17.

Glassman, D. M., and R. M. Crow. 1996. Standardization model for describing extent of burn injury to human remains. *Journal of Forensic Sciences* 41:152–154.

Goff, M. L. 1991. Comparison of insect species associated with decomposing remains recovered inside dwellings and outdoors on the island of Oahu, Hawaii. *Journal of Forensic Sciences* 36:748–753.

———1991. Feast of clues: Insects in the service of forensics. *The Sciences* 31:30–35.

———1992. Problems in estimation of postmortem interval resulting from wrapping of the corpse: A case study from Hawaii. *Journal of Agricultural Entomology* 9:237–243.

———1993. Estimation of postmortem interval using arthropod development and successional patterns. *Forensic Sciences Review* 5:81–94.

Goff, M. L., and M. M. Flynn. 1991. Determination of postmortem interval by arthropod succession: A case from the Hawaiian Islands. *Journal of Forensic Sciences* 36:607–614.

Goff, M. L., and W. D. Lord. 1994. Entomotoxicology: A new area for forensic investigation. *American Journal of Forensic Medicine and Pathology* 15:51–57.

Goff, M. L., and C. B. Odom. 1987. Forensic entomology in the Hawaiian Islands: Three case studies. *American Journal of Forensic Medicine and Pathology* 8: 45–50.

Goff, M. L., A. I. Omori, and K. Gunatilake. 1988. Estimation of postmortem interval by arthropod succession: Three case studies from the Hawaiian Islands. *American Journal of Forensic Medicine and Pathology* 9:220–225.

Goodbrod, J. R., and M. L. Goff. 1990. Effects of the larval population density on rates of development and interactions between two species of *Chrysomya* (Diptera: Calliphoridae) in laboratory culture. *Journal of Medical Entomology* 27:338–43.

Greenberg, B. 1971. *Flies and disease*, vols. 1 and 2. Princeton University Press, Princeton, N.J.

———1985. Forensic entomology: Case studies. *Bulletin of the Entomological Society of America* 31:25–28.

———1990. Blow fly nocturnal oviposition behavior. *Journal of Medical Entomology* 27:807–810.

———1991. Flies as forensic indicators. *Journal of Medical Entomology* 28:565–577.

Gunatilake, K., and M. L. Goff. 1989. Detection of organophosphate poisoning in a putrefying body by analyzing arthropod larvae. *Journal of Forensic Sciences* 34:714–716.

Harwood, R. F., and M. T. James. 1979. *Entomology in human and animal health,* 7th ed. Macmillan Publishing Co., New York.

Inoue, Y. 1964. Effects of several organophosphate insecticides against last instar maggots of the flesh fly. *Japanese Journal of Sanitary Zoology* 15:273–274.

Introna, F., C. LoDico, Y. H. Caplan, and J. E. Smialek. 1990. Opiate analysis in cadaveric blowfly larvae as an indicator of narcotic intoxication. *Journal of Forensic Sciences* 35:118–122.

Kamal, A. S. 1958. Comparative study of thirteen species of sarcosaprophagous Calliphoridae and Sarcophagidae (Diptera). *Annals of the Entomological Society of America* 51:261–271.

Kintz, P., A. Godelar, A. Tracqui, P. Mangin, A. A. Lugnier, and A. J. Chaumont. 1990. Fly larvae: A new toxicological method of investigation in forensic medicine. *Journal of Forensic Sciences* 35:204–207.

Lord, W. D., T. R. Adkins, and E. P. Catts. 1992. The use of *Synthesiomyia nudiseta* (Van Der Wulp) (Diptera: Muscidae) and *Calliphora vicina* (Robineau-Desvoidy) (Diptera: Calliphoridae) to estimate the time of death of a body buried under a house. *Journal of Agricultural Entomology* 9:227–235.

Lord, W. D., M. L. Goff, T. R. Adkins, and N. H. Haskell. 1994. The black soldier fly *Hermetia illucens* (Diptera: Stratiomyidae) as a potential measure of human postmortem interval: Observations and case studies. *Journal of Forensic Sciences* 39:215–222.

Lord, W. D., and W. C. Rodriguez III. 1989. Forensic entomology: The use of insects in the investigation of homicide and untimely death. *The Prosecutor* Winter:41–48.

Lord, W. D., and J. R. Stevenson. 1986. *Directory of forensic entomologists,* 2nd ed. Defense Pest Management Information Analysis Center, Walter Reed Army Medical Center, Washington, D.C.

McKnight, B. E. 1981. *The washing away of wrongs: Forensic medicine in thirteenth-century China.* Ann Arbor: University of Michigan Press.

Megnin, J. P. 1894. La faune des cadavers: Application e l'entomologie a la médecine légale. Encyclopédie Scientifique des Aide-Mémoire. Maison Gauthiers-Villas et Fils, Paris.

Nuorteva, P. 1977. Sarcosaprophagous insects as forensic indicators. In *Forensic medicine: A study in trauma and environmental hazards,* vol. 2., ed. C. G. Tedeschi, W. C. Eckert, and L. G. Tedeschi, pp. 1072–1095. Saunders, Philadelphia.

Payne, J. A. 1965. A summer carrion study of the baby pig *Sus scrofa* Linnaeus. *Ecology* 46:592–602.

Redi, F. 1668. *Esperienze intorno alla generazione degli insetti.* Insegna della Stella, Florence.

Reed, H. B. 1958. A study of dog carcass communities in Tennessee, with special reference to the insects. *American Midland Naturalist* 59:213–245.

Richards, E. N., and M. L. Goff. 1997. Arthropod succession on exposed carrion in three contrasting tropical habitats on Hawaii Island, Hawaii. *Journal of Medical Entomology* 34:328–339.

Rodriguez, W. C., and W. M. Bass. 1983. Insect activity and its relationship to decay rates of human cadavers in east Tennessee. *Journal of Forensic Science* 28:423–432.

———1985. Decomposition of buried bodies and methods that may aid in their location. *Journal of Forensic Sciences* 30:836–852.

Schoenly, K., M. L. Goff, and M. Early. 1992. A BASIC algorithm for calculating the postmortem interval from arthropod successional data. *Journal of Forensic Sciences* 37:808–823.

Schoenly, K., M. L. Goff, J. D. Wells, and W. D. Lord. Quantifying statistical uncertainty in succession-based entomological estimates of the postmortem interval in death scene investigations: A simulation study. *American Entomologist* 42:106–112.

Smith, K. G. V. 1986. *A manual of forensic entomology.* British Museum (Natural History), London.

Tullis, K., and M. L. Goff. 1987. Arthropod succession in exposed carrion in a tropical rainforest on O'ahu Island, Hawai'i. *Journal of Medical Entomology* 24:332–339.

Webb, J. P., Jr., R. B. Loomis, M. B. Madoon, S. G. Bennett, and G. E. Green. 1983. The chigger species *Eutrombicula belkini* Gould (Acari: Trombiculidae) as a forensic tool in a homicide investigation in Ventura County, California. *Bulletin of the Society* of *Vector Ecology* 8:141–146.

Zumpt, F. 1965. *Myiasis in man and animals in the Old World.* Butterworths, London.

Acknowledgments

I am indebted to a number of people for help in the preparation of this book. I have had the unfailing support of my wife, Dianne, and my daughters, Dana and Alaina. They have tolerated and supported my efforts in a field which is not all that easy to explain. For the past 16 years, I have been involved in the evolution of forensic entomology from a seldom used novelty to its current status as a recognized tool in criminal investigations. While in this book I am presenting my own experiences, I have certainly not been the only person participating in this evolution. I have been fortunate in having the friendship and support of a number of others working in forensic entomology and related areas, in particular Wayne D. Lord, William C. Rodriguez III, Edward McDonough, Alvin I. Omori, Kanthi Von Guenthner, Wilson Sullivan, and Gary Dias. Without their encouragement, much of this work would not have been possible. I am grateful to my editors, Ann Downer-Hazell and Nancy Clemente, Ann for suggesting this project and both of them for reminding me how to write in complete English sentences. The excellent illustrations are by Amy Bartlett Wright. The College of Tropical Agriculture and Human Resources, University of Hawaii at Manoa, has been generous in allowing and encouraging my investigations of a field that is, at best, only marginally related to agriculture. A portion of the proceeds from *A Fly for the Prosecution* will go to Helping Hands Hawaii to assist in their efforts on behalf of victims and survivors of crimes.

Index

Page numbers in ***boldface*** *refer to illustrations*

www.smpl.org